Springer Desktop Editions in Chemistry

L. Brandsma, S. F. Vasilevsky, H. D. Verkruijsse
Application of Transition Metal Catalysts in Organic Synthesis
ISBN 3-540-65550-6

H. Driguez, J. Thiem (Eds.)
Glycoscience, Synthesis of Oligosaccharides and Glycoconjugates
ISBN 3-540-65557-3

H. Driguez, J. Thiem (Eds.)
Glycoscience, Synthesis of Substrate Analogs and Mimetics
ISBN 3-540-65546-8

H. A. O. Hill, P. J. Sadler, A. J. Thomson (Eds.)
Metal Sites in Proteins and Models, Phosphatases, Lewis Acids and Vanadium
ISBN 3-540-65552-2

H. A. O. Hill, P. J. Sadler, A. J. Thomson (Eds.)
Metal Sites in Proteins and Models, Iron Centres
ISBN 3-540-65553-0

H. A. O. Hill, P. J. Sadler, A. J. Thomson (Eds.)
Metal Sites in Proteins and Models, Redox Centres
ISBN 3-540-65556-5

A. Manz, H. Becker (Eds.)
Microsystem Technology in Chemistry and Life Sciences
ISBN 3-540-65555-7

P. Metz (Ed.)
Stereoselective Heterocyclic Synthesis
ISBN 3-540-65554-9

H. Pasch, B. Trathnigg
HPLC of Polymers
ISBN 3-540-65551-4

T. Scheper (Ed.)
New Enzymes for Organic Synthesis, Screening, Supply and Engineering
ISBN 3-540-65549-2

Springer
Berlin
Heidelberg
New York
Barcelona
Hong Kong
London
Milan
Paris
Singapore
Tokyo

T. Scheper (Ed.)

New Enzymes for Organic Synthesis

Screening, Supply and Engineering

 Springer

Prof. Dr. Thomas Scheper
Universität Hannover
Institut für Technische Chemie
Callinstraße 3
D-30167 Hannover
Germany

Description of the Series

The Springer Desktop Editions on Chemistry is a Paperback series that offers selected thematic volumes from Springer chemistry series to graduate students and individual scientists in industry and academia at very affordable prices. Each volume presents an area of high current interest to a broad non-specialist audience, starting at the graduate student level.

Formerly published as hardcover edition in the review series
Advances in Biochemical Engineering/Biotechnolgy (Vol. 58) ISBN 3-540-61689-6

Cataloging-in-Publication Data applied for

ISBN 3-540-65549-2
Springer-Verlag Berlin Heidelberg New York

Cover: design & production, Heidelberg
Typesetting: Macmillan India Ltd. Bangalore-25
SPIN: 10711899   02/3020 -5 4 3 2 1 0 - Printed on acid-free paper

Professor M.-R. Kula

Preface

New enzymes for organic synthesis continue to represent a challenge both for those who prepare them as well as for those who finally apply them. The improved accessibility of enzymes as industrial catalysts has led to a large number of practical processes. The times are long past when it was believed that, at best, hydrolases would be suitable for industrial operations. Today, there are examples of practical applications for almost all enzyme classes.

Professor Dr. Maria-Regina Kula (Düsseldorf and Jülich, Germany) is a Pioneer in this field and it is therefore quite logical that this volume should appear on the occasion of her sixtieth birthday.

Professor Kula spent six month on sabbatical leave at the California Institute of Technology with Professor Frances H. Arnold, who has contributed a paper on the topic of "Optimizing Industrial Enzymes by Directed Evolution". Professor Arnold shows how enzymes usable in practice, for which in nature under "natural conditions" there has been no evolutionary stress, can result from directed evolution.

Profesor Kula has also been scientifically associated for many years with Professor Sakayu Shimizu, Department of Agricultural Chemistry, Kyoto University, Japan. Together with him and his predecessor, Professor Hideaki Yamada, she has organized the German-Japanese workshops on enzyme technology for many years. Professor Shimizu writes about "Screening of Novel Microbial Enzymes for the Production of Biologically and Chemically Useful Compounds". Many practical processes have originated from the Department of Agricultural Chemistry at Kyoto University and the technology transfer of this institute in particular is exemplary.

The other four chapters of this volume are contributed by heads of department from the Institute of Enzyme Technology at Heinrich Heine University, Düsseldorf, of which Prof. Kula is the director. First of all, Privatdozent Dr. Werner Hummel concerns himself with "New Alcohol Dehydrogenases for the Synthesis of Chiral Compounds". Up to just a few years ago, it was believed that due to the cofactor regeneration problem oxidoreductases could not be applied in practice. In his contribution, Dr. Hummel shows that an interesting pathway to chiral alcohols has been opened up by the provision of sufficient quantities of alcohol dehydrogenases. These enzymes are used together with formate dehydrogenase and formate as a hydrogen source. Credit for the fact that formate dehydrogenase can today be termed an industrial biocatalyst is undoubtedly due to Professor Kula.

In the next paper, Privatdozent Dr. Lothar Elling is concerned with "Glycobiotechnology: Enzymes for the Synthesis of Nucleotide Sugars". Dr. Elling shows it is also true of glycobiotechnology that this field can only develop if corresponding enzymes are made available in sufficient quantities. Enzyme-catalysed oligosaccharide synthesis will undoubtedly receive great impetus from the inexpensive preparation of activated sugars.

Dr. Martina Pohl has entitled her contribution "Protein Design on Pyruvate Decarboxylase (PDC) by Site-Directed Mutagenesis". New enzymes cannot only be obtained by directed evolution but also by site-directed mutagenesis so that very interesting complementary approaches emerge here. Particularly in the past view years, success has been achieved in selectively modifying the substrate spectrum and reaction conditions for enzymes used in practical operations.

Finally, in a concluding paper Dr. Jörg Thömmes considers enzyme recovery in "Fluidized Bed Adsorption as a Primary Recovery Step in Protein Purification". As important as it is to track down new enzymes and selectively modify them, it remains equally important to actually make them available in the flask on the bench in adequate quantities at low cost with sufficient purity. Recovery is of central significance in this respect. Fluidized bed adsorption combines the process steps of cell separation, concentration and primary cleaning in recovery work. The procedure can also be excellently transferred from the laboratory to the pilot scale.

Professor Kula has performed really outstanding work in all these fields. Many of her collegues may be unaware that the subject of her doctoral dissertation concerned inorganic chemistry. She first came into intensive contact with enzymes during her postdoc time in the School of Medicine at John Hopkins University, USA. She later became head of a department and finally scientific director (1975-1979) at the Gesellschaft für Biotechnologische Forschung mbH (GBF), in Braunschweig, Germany.

To date she is, at least in Germany, the only woman ever to have been in charge of a national laboratory.

Since 1986 she has been engaged as a professor at the University of Düsseldorf and at the Research Centre Jülich. She has received numerous honours (including Order of Merit of the Federal Republic of Germany in 1979, the Technology Transfer Prize of the German Research Minister in 1983, and the Enzyme Engineering Award, USA, in 1995).

Preface

As a long-standing cooperation partner of Professor Kula, I feel the need to express my thanks as a kind of representative of the scientific community on the occasion of her sixtieth birthday for her numerous, fascinating scientific contributions and her exemplary willingness to cooperate.

Biotechnolgy is the integrated application of different scientific disciplines. Professor Kula has always exemplified this together with her staff. On the basis of long-term scientific work, real practical innovations have emerged due to her own qualities and her receptiveness for other scientific fields. Only a few poeple who have cooperated with her as long as I have will be aware of how much she has always enjoyed her work and the humour which accompanies her successes and setbacks.

We wish her continuing enthusiasm for her work and for us all more "new enzymes for organic synthesis".

Jülich, January 1997   Prof. Dr. C. Wandrey

Contents

Optimizing Industrial Enzymes by Directed Evolution

Frances H. Arnold and Jeffrey C. Moore
Division of Chemistry and Chemical Engineering 210-41,
California Institute of Technology, Pasadena, CA 91125, USA

Dedicated to Professor Dr. Maria-Regina Kula on the occcasion of her 60th birthday

Enzymes can be tailored for optimal performance in industrial applications by directing their evolution in vitro. This approach is particularly attractive for engineering industrial enzymes. We have created an efficient para-nitrobenzyl esterase over six generations of random point mutagenesis and recombination coupled with screening for improved variants. The best clones identified after four generations of sequential random mutagenesis and two generations of random recombination display more than 150 times the p-nitrobenzyl esterase activity of wild type towards loracarbef-p-nitrobenzyl ester in 15% dimethylformamide. Although the contributions of individual effective amino acid substitutions to enhanced activity are small (< 2-fold increases), the accumulation of multiple mutations by directed evolution allows significant improvement of the biocatalyst for reactions on substrates and under conditions not already optimized in nature. The positions of the effective amino acid substitutions have been identified in a pNB esterase structural model. None appear to interact directly with the antibiotic substrate, further underscoring the difficulty of predicting their effects in a 'rational' design effort.

Advances in Biochemical Engineering/
Biotechnology, Vol. 58
Managing Editor: Th. Scheper
© Springer-Verlag Berlin Heidelberg 1997

1 Introduction

Enzyme processes are making significant inroads in the production of foods, chemicals, and pharmaceuticals, thanks in large part to the sustained efforts of researchers like Professor Kula and her colleagues, who have identified enzymes to catalyze key transformations and developed viable processes to accompany them. Although the opportunities for using enzymes are numerous, identification of appropriate enzyme catalysts remains one of the key limiting steps.

Chemical engineers who design industrial processes using enzymes are constantly stymied by the fact that these catalysts have evolved over billions of years to perform very specific biological functions and to do so within the context of a living organism. Some of the features required for function in a complex chemical network are undesirable when the catalyst is lifted out of context (e.g. product inhibition). Conversely, many of the properties we wish an enzyme would have clash with the needs of the organism, or at least were never required: high stability, the ability to function in nonnatural environments and catalyze nonnatural reactions.

Despite intense research, there are enormous gaps in our understanding of the relationships between amino acid sequence, structure and function. As a result, the rational design of new proteins by the classical reductionist approach can be a frustrating, and often fruitless, exercise. Clues as to how to engineer better enzymes, however, come from studying how nature has done it. A study of protein evolution shows that enzymes are highly adaptable molecules, at least over evolutionary time scales. Many enzymes catalyzing very different reactions have come about by divergent evolution from a common ancestral protein of the same general structure, acquiring diverse capabilities by processes of random mutation, recombination, and natural selection. We also know that enzymes of a given function (for example, all catalyzing a particular step in a metabolic pathway) can exhibit widely different properties (stability, solubility, tolerance to pH, etc.), depending on where they are found.

The explosion of tools that has come out of molecular biology during the last 20 years has made it possible for us to evolve enzymes for features never required in nature. We can speed up the rate and channel the direction of evolution by controlling mutagenesis and the accompanying 'selection' pressures. By uncoupling the enzymes from the constraints of function within a living system, we can explore a variety of futures that include novel environments or even entirely new functions. Directed evolution is also a very practical approach to tailoring enzymes for a wide range of applications. Our recent review [1] describes just some of the protein properties that have been successfully altered by evolutionary approaches: substrate specificity, catalytic activity, activity in the presence of organic solvents, expression level and stability.

The methods are especially suitable for optimizing industrial enzymes. For example, we have recently directed the evolution of an enzyme to catalyze a nonnatural reaction, *p*-nitrobenzyl (pNB) ester hydrolysis in the presence of polar organic solvents [2]. Scientists at a major pharmaceutical company, Eli Lilly, devoted significant effort to finding an enzyme that would selectively remove the pNB protecting group commonly used during large-scale synthesis of cephalosporin-type antibiotics [3, 4]. An enzyme with some pNB esterase activity was identified in *B. subtilis*, but its low activity, especially in the organic solvents required to solubilize these materials, made it a poor competitor to the existing zinc catalyst. We successfully evolved this enzyme to exhibit much higher activity towards the loracarbef pNB ester (LCN-pNB), both in aqueous buffer and in buffer-dimethylformamide solutions.

2 A Working Strategy for Directed Enzyme Evolution

The number of possible enzymes one can make is so vast than an exploration of their functions must be carefully guided in order to avoid becoming hopelessly lost. It is therefore important to develop a practical working strategy that will guide evolution in the laboratory. In nature, unfavorable mutations are winnowed out at the same time as beneficial mutations are amplified, by linking the organism's growth rate and reproductive success to the performance of its components. In this process of *selection*, those organisms which grow faster quickly dominate, allowing an efficient search of very large populations (10^6 and more for bacteria). Unfortunately, many of the features that are of interest for applications cannot be linked to the survival or growth of the host organism – the prerequisite to a selection. Therefore, mutant enzyme libraries must be screened. This unfortunate reality, a direct result of the first law of random mutagenesis ("You get what you screen for"), effectively limits the search for improvements to mutant libraries containing perhaps 10^4–10^6 variants [5].

The number of different sequences one can create by making mutations in an enzyme grows exponentially with the number of mutations. While there are only 5700 possible single mutants of a 300 amino acid enzyme, there are more than 30 billion different sequences that differ from the original enzyme at only three positions. While a rapid screen might be able to cover a large fraction of all single mutants, and even some significant fraction of all double mutants, screening would be unable to give more than a very sparse sampling of the enzymes with multiple amino acid substitutions. Beneficial mutations are generally rare; their frequency will depend on the extent to which the particular feature of interest has already been optimized. Because mutations are more likely to be harmful than beneficial, the probability that an enzyme will be improved with respect to its parent sequence decreases rapidly with increasing mutation rate. As a result, the search for effective mutations should be limited to proteins with sequences (and therefore properties) very similar to their parents. We try to tune the rate of mutation to produce enzyme libraries with primarily single (amino acid) substitutions.

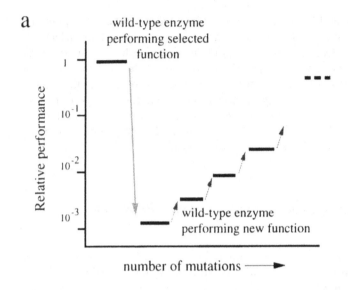

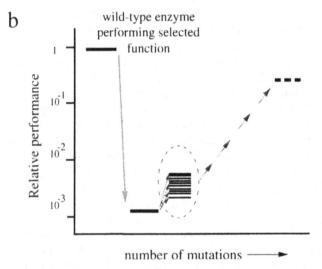

Fig. 1. A working strategy for directed enzyme evolution. The screening method should ensure that small enhancements brought about mainly by single mutations can be measured. The evolution of a new, useful enzyme requires an effective strategy for accumulating many such small improvements. Beneficial mutations can be accumulated in **a** sequential generations of random mutagenesis and screening or **b** by (random) recombination

Based on these arguments, we developed the simple directed evolution strategy illustrated in Fig. 1 [2, 6]. In comparison to the enzyme performing a function for which it is selected in nature, the new job is performed poorly indeed. Beneficial mutations will be identified during screening the products of gene libraries containing relatively few (single or perhaps double) amino acid substitutions. Although these progeny will generally resemble their parents, the accumulation of mutations creates descendents quite different from their ancestor. Therefore, the generation of new, useful enzymes also relies on having an effective strategy for accumulating many such small improvements. One such strategy, illustrated in Fig. 1a, involves carrying out sequential generations of random mutagenesis to create a mutant library, coupled with screening of the resulting enzyme variants. In each generation, a single variant is chosen to parent the next generation, and sequential cycles allow the evolution of the desired features. Alternatively, effective mutations identified during one or more generations can be recombined, for example using the recently developed 'DNA shuffling' method described by Stemmer [7, 8] (Fig. 1b). Both approaches to accumulating beneficial mutations have proven effective for the evolution of the pNB esterase.

3 Screening for Improved Enzymes

The first requirement for successful laboratory evolution is the development of a sensitive screen to identify improved enzymes. The screen should ensure that the expected small enhancements brought about mainly by single amino acid substitutions can be measured. While a carefully-designed selection may be useful for making a first pass (i.e. to remove a large background of inactive clones [9]), most industrial enzyme systems will require screening to identify useful mutations. The resulting evolved enzymes will not be useful unless they exhibit a combination of features: high expression levels, stability and enantioselectivity, high activity on nonnatural substrates and/or the ability to carry out the reaction at a particular temperature, pH, substrate concentration, in organic solvents, etc. These features are unlikely to be reflected in an in vivo selection. Furthermore, if the screen does not reflect all desired criteria, one runs the risk of evolving one property at the expense of those not required to pass the screen.

If two properties are closely linked to one another, however, the screen may need only reflect one. Calculated risks of this type are almost always necessary in designing a screen suitable for dealing with large numbers of variants. The p-nitrobenzyl ester hydrolysis reaction, for example, is assayed by HPLC, which is unsuitable for rapid screening. We therefore devised a screening assay based on the loracarbef p-nitrophenyl ester, which provides a colorimetric signal upon hydrolysis [2]. The screening reactions could then be carried out in a microtiter plate, using a plate reader to analyze the enzyme kinetics of all wells simultaneously. To validate the screening method, we compared the activities of a set of mutants towards the p-nitrobenzyl and p-nitrophenyl substrates, as shown in Fig. 2. If the screening reaction perfectly mimicked the desired reaction, all the

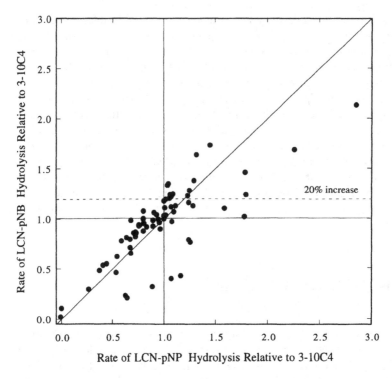

Fig. 2. Comparison of activities on target (*p*-nitrobenzyl) and screening (*p*-nitrophenyl) substrate of selected active pNB esterases isolated after fourth generation of random mutagenesis and screening, relative to parent enzyme from the 3rd generation. The five most active variants were pooled for random recombination

points would lie on the 45° line. Although there will be some false positives and false negatives using this screening reaction, the rapid screen provides sufficient information to make a rough cut of positive clones. Thus we need only test a small number of clones by HPLC to verify improved pNB esterase activity. Screening reactions were carried out at room temperature, in 15–20% DMF, with relatively high substrate concentrations (0.8 mM) to reflect the conditions under which the enzyme would be used industrially.

4 Evolution of pNB Esterase: Random Mutagenesis

Using the rapid colorimetric assay to screen about a thousand colonies per generation, we completed four generations of random mutagenesis (by error-prone PCR) and screening the *E. coli* colonies expressing the resulting gene libraries. The detailed methods and results have been described [2]; results are

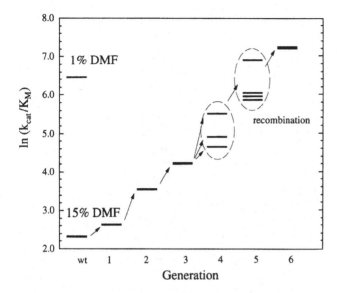

Fig. 3. Evolutionary progression of catalytic efficiency of pNB esterase (towards LCN-pNB) in 15% DMF through four generations of random mutagenesis, followed by two rounds of recombination by 'DNA shuffling' of circled populations.

summarized in Fig. 3. After four generations, the enzyme's catalytic efficiency in 15% DMF had improved 24-fold. Although not reflected in Fig. 3, expression level had also increased 2-fold.

The evolved enzymes were further characterized for their ability to carry out the desired pNB ester hydrolysis. Figure 4 shows the specific reaction rates for enzymes from the first four generations in 1% and 15% DMF. Each successive generation catalyst is more effective than its parent, and the best, pNB esterase 4-54B9, is 15 times more productive than wild type in 1% DMF. In 15% DMF, this enzyme makes product at 4 times the rate of the wild type enzyme in 1% organic solvent. The impact of this improvement is not only the increased productivity of the evolved enzyme, but also in the 4-fold increase in solubility of the substrate in 15% DMF. The increased solubility reduces the size of the reactor and the downstream processes required to produce and purify a given amount of product. The 2-fold increase in enzyme expression level further reduces process costs.

5 Evolution of pNB Esterase: In Vitro Recombination

In the fourth generation, we screened a larger number of clones ($\sim$7500) and picked a subset of those to analyze further, in order to validate the screening

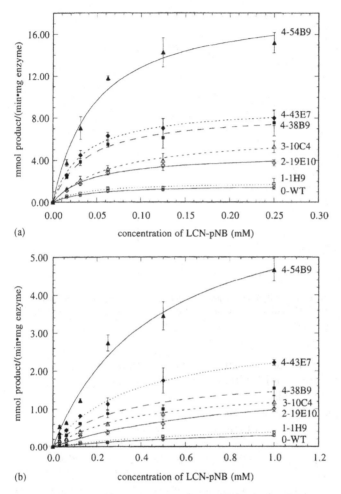

Fig. 4. Reaction rates for evolved pNB esterases from the first four generations of directed evolution. Activity towards *p*-nitrobenzyl loracarbef nucleus in 1% (**a**) and 15% (**b**) DMF in 0.1 M PIPES, pH 7.0, 30°C. Four times as much substrate can be dissolved in 15% DMF as in 1% DMF

method (Fig. 2). At this point we had a set of five confirmed positive pNB esterase variants (more than 50% more active than the parent from the 3rd generation) with which we could test the alternate, recombination approach for accumulating effective mutations (Fig. 1b).

Recombination is a powerful force in natural evolution: it serves to propagate positive traits and eliminate negative traits in progeny. Stemmer has described a method for random gene recombination in vitro, which he calls 'DNA shuffling' or 'sexual PCR' [7.8]. Initially, the DNA from a collection of enzyme variants to be shuffled is amplified by PCR. The resulting DNA is randomly digested into a pool of small fragments by DNase I. A PCR-based reassembly step takes advantage of the overlapping regions of sequence identity between adjacent fragments to direct in primer/template fashion the synthesis of complete sequences. This step performs the recombination, as fragments with and without mutations serve as templates for other priming fragments containing or lacking mutations. A final PCR amplification step provides large quantities of the recombinant pool for cloning.

In vitro recombination has several useful features. First, it makes use of existing beneficial mutations. Screening a large number of pNB esterase clones in the fourth generation identified five significantly more active than the parent. Choosing any one of these for another cycle of random mutagenesis and screening would have meant losing the information present in the other four. When combined with screening, DNA shuffling allows the discovery of the best combinations of mutations and does not assume that the best combination contains all the mutations in a population.

Second, recombination occurs simultaneously with point mutagenesis. An effect of forcing DNA polymerase to synthesize full length genes from the small fragment DNA pool is a background mutagenesis rate, which can be controlled. Where selection can be employed to search the resulting libraries, such as the laboratory evolution of antibiotic resistance, huge progress has been made in evolving enzymatic activity, as represented by a 16 000 fold increase in the amount of antibiotic required to inhibit bacterial growth [7]. This is a direct result of the extraordinarily large diversity this process can create and natural selection can search (10^6 or more different bacteria per experiment). The background mutagenesis yielded the genetic variability on which recombination acted to enhance antibiotic resistance.

A third feature of recombination is that it can be used to remove deleterious mutations. Positive variants containing multiple mutations in most cases contain one mutation responsible for improved behavior and one or more extraneous mutations which are neutral or even slightly deleterious. These extraneous mutations can negatively affect properties not explicitly required under the screening or selection conditions. For example, pNB esterase was screened for increased activity on p-nitrophenyl loracarbef nucleus (LCN-pNP) in dimethylformamide (DMF). The screening was performed at 30 °C. The wild type enzyme denatures at temperatures above 55 °C, and we can imagine that enhanced activity variants containing multiple extraneous mutations could be less thermostable. DNA shuffling with an excess of wild-type DNA (back-crossing) can remove extraneous mutations which survive the primary screen [8].

Finally, recombination has the advantage of allowing parallel processing in screening. Several screens could be processed simultaneously for searching a variety of different criteria. Mutations which provide enhanced activity could be examined in one screen, while other screens could examine improvements in stability, solvent tolerance, expression levels or other properties. Positives could then be recombined to search for variants improved in several properties at

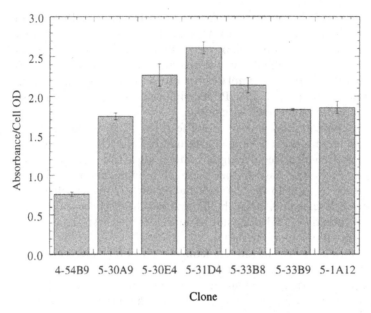

Fig. 5. Activities of positive clones isolated from library produced by shuffling five fourth-generation pNB esterase genes. Clone 4-54B9 is the most active parent. 5-1A12 was identified after restriction/ligation of 4-54B9 and 4-38B9

once. A final consideration is that recombination can save work in fine-tuning the screening conditions. Screening conditions can be difficult to optimize, especially when the variants are screened under conditions where enzymatic properties change dramatically, such as near the melting temperature. The ability to recombine mutations allows us to identify several beneficial mutations at one well-researched set of conditions. Of course, recombination works when the effects of individual mutations are at least cumulative, if not perfectly additive. Our experience, however, is that small effects are very often cumulative, even for properties not generally recognized to be highly distributed, such as catalytic activity or substrate specificity [6].

The genes from the five improved fourth generation pNB esterases were shuffled using Stemmer's method [10]. From the resulting library of pNB esterase recombinants, five variants demonstrating activity greater than 4-54B9 were identified after screening 440 colonies. The activities of these five variants and the fifth generation variant produced by a simple restriction/ligation shuffling (5-1A12, [2]) are compared to 4-54B9 in Fig. 5. All but one of the screened variants display activity comparable to 5-1A12. A clone expressing variant 5-31D4 demonstrates a 3.5-fold increase in activity over 4-54B9, for more than a 100-fold increase in total activity over the clone expressing wild type pNB esterase.

Sequencing revealed that the original five variant (fourth generation) pool contained only four unique variants. These four were reshuffled and rescreened, with results very similar to those in Fig. 5. Six positive clones were identified, four of which were approximately two-fold more active than 4-54B9, and one of which was four-fold more active. These six clones were then shuffled together, and screened to identify two clones with more than a 5.5-fold increase in activity over 4-54B9. The overall results of two cycles of shuffling and screening approximately 2000 additional colonies are summarized in Fig. 2. The best pNB esterase clone exhibits more than 150-fold greater total activity than the wild-type clone. Recombination of the pNB esterases led to even more highly evolved enzyme variants. Thus, by DNA shuffling, we were able to make use of the information available in the collection of improved fourth-generation variants.

6 Sequence and Structural Analysis

All the evolution experiments described here can be performed in the absence of sequence information. The sequences of the evolved pNB esterases with demonstrated improvements in activity have nonetheless been determined. A lineage of all the pNB esterase variants sequenced to date showing the amino acid substitutions obtained during each generation of directed evolution is provided in Fig. 6.

The first four cycles of random mutagenesis and screening led to the generation and accumulation of amino acid substitutions. In each generation, one (amino acid) mutation is expected to lead to the enhanced activity. Where two mutations appear in a single generation, we have argued that one is likely to be neutral or slightly deleterious [2]. The restriction/ligation experiment that generated 5-1A12 in the 5th generation involves three mutations: I60V in one parent (4-38B9) and K267R and L334S in the other parent (4-54B9). Restriction/ligation occurred between the mutations leading to K267R and L334S, yielding a more active enzyme with I60V and L334S present and K267R removed. Thus in this simple recombination experiment, we saw both the joining of positive mutations and the elimination of a neutral (or negative) mutation.

The first DNA shuffling experiment was done before the sequences had been determined. Two of the shuffled variants, 4-38B9 and 4-73B4, turned out to be identical. Furthermore, it was learned that 4-54B9 and 4-43E7 share different mutations in the same codon. This reduces the number of available positions for recombination from five to only three. Figure 6 lists the amino acid substitutions found in 5-1A12 and the two most active variants identified in the first round of DNA shuffling. All three contain the mutations which lead to the Ile 60 Val substitution from 4-38B9 and Leu 334 Ser from 4-54B9. Variant 5-30E4 appears 25% more active than 5-1A12; this may be due to an increase in specific activity as a result of the new translated mutation, Ala 454 Val. These enzymes have not yet been purified and characterized. Therefore, the observed increase in activity could also reflect an increase in total functional protein produced as a result of

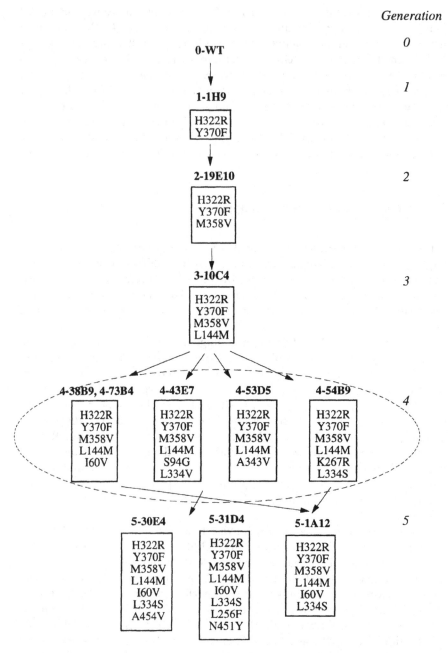

Generation

Fig. 6. Lineage of pNB esterase variants generated by directed evolution showing the amino acid substitutions found in each

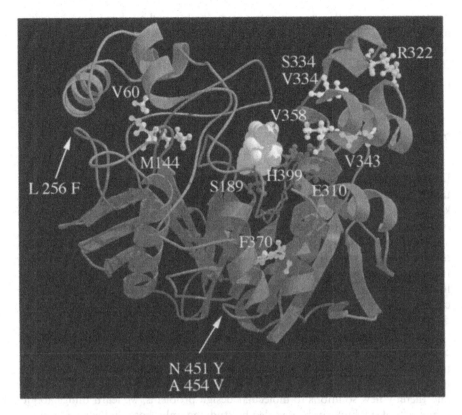

Fig. 7. Molecular model of the pNB esterase showing positions of antibiotic *p*-nitrobenzyl ester substrate (white CPK structure), catalytic residues (S189, E310, and H399), and beneficial mutations accumulated during directed evolution. Mutations at positions 322 and 370 are believed to improve expression, while the remaining six substitutions improve specific activity [2]. Arrows indicate the position of new mutations found after DNA shuffling

any of the several changes at the DNA level which are silent at the amino acid level. Variant 5-31D4 is 50% more active than 5-1A12 and contains 5 new mutations, of which 2 are translated. Again the increased activity observed for this clone could reflect an increase in specific activity or total protein.

The positions of the amino acid substitutions identified in the various pNB esterases are illustrated in Fig. 7, on a model of the pNB esterase developed from the X-ray crystal structures of homologous enzymes [2]. Known beneficial mutations are indicated; also shown are the mutations believed to be neutral or to affect expression. Positions of the additional translated mutations found in the variants produced by DNA shuffling are indicated by the white arrows. None of the effective amino acid substitutions lie in segments of the esterase predicted to interact directly with the bound substrate. It may be that the amino acid substitutions sampled at positions adjacent to the substrate were all deleterious, and small improvements were only obtained by altering amino acids further away. The effects of individual mutations are small, and the mechanism(s) by which these amino acid substitutions enhance the catalytic activity of the evolved pNB esterases are subtle. These would have been impossible to predict in advance.

7 Conclusions

Directed evolution offers a unique opportunity for biotechnology: the ability to tailor proteins for optimal performance in a wide range of applications. We know that the enzymes isolated and characterized to date represent a tiny fraction of the functions in the huge biological bank displayed by nature. Limiting ourselves even to this much larger set is shortsighted, however, as these molecules, too, have evolved in living organisms and therefore suffer many limitations. The *possible* function space is much larger indeed, and this space is likely to be of the most interest for applications. Directed evolution offers us the chance to explore this vast space of potentially useful catalysts, free from the constraints of function within a living system.

In contrast to 'rational' design approaches, directed evolution can be applied even when very little is known about an enzyme's structure or catalytic mechanism. Since the vast majority of proteins remain largely uncharacterized, this marks a huge advantage for the evolutionary methods. By evolving new functions and thereby new solutions to molecular design problems, we learn things about enzymes that might never be revealed by studying only those that exist in nature.

Acknowledgments. This work was supported by the U.S. Office of Naval Research and the U.S. Department of Energy's program in Biological and Chemical Technologies Research within the Office of Industrial Technologies, Energy Efficiency and Renewables.

8 References

1. Shao Z, Arnold FH (1996) Curr Opin Struct Biol 6: 513
2. Moore J, Arnold FH (1996) Nature Biotechnol 14: 458
3. Brannon DR, Mabe JA, Fukuda DS (1976) J Antibiotics 29: 121
4. Zock J, Cantwell C, Swartling J, Hodges R, Pohl T, Sutton K, Rosteck Jr. P, McGilvray D, Queener S (1994) Gene 151: 37
5. You L, Arnold FH (1996) Prot Eng 9: 77
6. Chen K, Arnold FH (1993) Proc Natl Acad Sci USA 90: 5618
7. Stemmer WPC (1994) Proc Natl Acad Sci USA 91: 10747
8. Stemmer WPC (1994) Nature (Lond) 340: 389
9. Black ME, Newcomb TG, Wilson HMP, Loeb LA (1996) Proc Natl Acad Sci USA 93: 3525
10. Moore JC, Jin HM, Arnold FH (in preparation)

Protein Design on Pyruvate Decarboxylase (PDC) by Site-Directed Mutagenesis

Application to Mechanistical Investigations, and Tailoring PDC for the Use in Organic Synthesis

Martina Pohl
Institut für Enzymtechnologie, der Heinrich-Heine Universität Düsseldorf, im Forschungszentrum Jülich, Postfach 2050, D-52402 Jülich

Dedicated to Professor Dr. Maria-Regina Kula on the occasion of her 60th birthday

Pyruvate decarboxylases (E.C. 4.1.1.1) from various organisms have been studied for many years, mainly with respect to the mechanism of the non-oxidative decarboxylation reaction. Although the C–C-bond-forming properties of these enzymes are known and have been applied for many years in biotransformations for the synthesis of chiral α-hydroxy ketones, only little is known about the factors influencing the carboligase side-reaction.

The present review surveys recent efforts in the study of site-directed mutagenesis on PDCs, which are discussed against a background of the structural and kinetical investigations. It also includes recent studies on tailoring the PDCs of *Zymomonas mobilis* for the syntheses of (*R*)-phenylacetyl carbinol (PAC), a pre-step in the synthesis of L-ephedrine, by protein design techniques.

Advances in Biochemical Engineering/
Biotechnology, Vol. 58
Managing Editor: Th. Scheper
© Springer-Verlag Berlin Heidelberg 1997

List of Symbols and Abbreviations

ADH	alcohol dehydrogenase
CD	circular dichroism
3D	three-dimensional
HEThDP	2-(hydroxyethyl)thiamine diphosphate
Mes	4-morpholineethanesulfonsäure
n_h	Hill-coefficient
PAC	phenylacetyl carbinol
PDC	pyruvate decarboxylase
PDC$S.c.$	PDC from *Saccharomyces cerevisiae*
PDC$S.u.$	PDC from *Saccharomyces uvarum*
PDC$Z.m.$	PDC from *Zymomonas mobilis*
$S_{0.5}$	substrate concentration necessary for half-maximal velocity
ThDP	thiamine diphosphate
v/S	velocity vs substrate concentration
V_{max}	maximal velocity
wt	wild-type

1 Introduction

Pyruvate decarboxylases combine two different synthetic properties: the non-oxidative decarboxylation of α-keto acids and a carboligase side-reaction leading to the formation of α-hydroxy ketones. The decarboxylation of pyruvate to acetaldehyde by fermenting yeast was first described by Neuberg and coworkers [1–3]. Neuberg and Hirsch detected the carboligase activity in 1921 during their studies of "phytochemical reductions" of various aldehydes by fermenting yeast [4]. This reaction was studied in detail by analyzing the reaction products which could be obtained by the reduction of benzaldehyde. Besides the expected benzyl alcohol, a new optically active compound with the molecular formula $C_9H_{10}O_2$ was detected and later characterized as (–)-phenylacetyl carbinol (PAC) ((R)-1-hydroxy-1-phenylpropan-2-one) **1** [5]. Neuberg named the new enzyme "carboligase", and assumed it to exist apart from "α-carboxylase" (PDC) in yeast.

The production of (R)-PAC by fermenting yeast was one of the first industrially applied biotransformations, which is used to nowadays to obtain (R)-PAC as a chiral pre-step for L-ephedrine **2** [6–9] (Scheme 1), and the optimization of the fermentation process is still a matter of research [10–23].

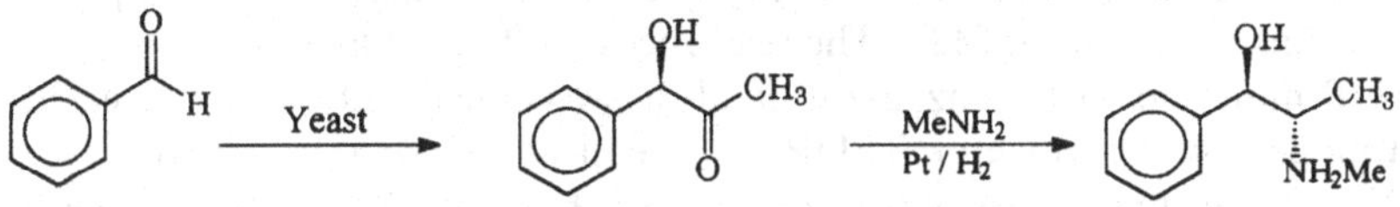

Scheme 1. Synthesis of (–)-ephedrine

In 1930 there were indications that the production of α-hydroxy ketones is a side-reaction of the PDC [24, 25]. These results were later confirmed by the work of Singer and Pensky [26, 27], who detected the production of acetoin by PDC from wheat germ using either acetaldehyde or pyruvate as precursors. Juni [28, 29] was unable to separate PDC from the acetoin-forming system of brewer's yeast, wheat germ and bacteria. Unequivocal proofs for PDC as the origin of both the carboligase and decarboxylase activity have been obtained by studying the mechanistical background of both reactions (see Sect. 2).

The importance of magnesium ions for the catalytic activity of the yeast PDC was recognized in 1931 by Lohmann and Schuster [30]. One year later, Auhagen [31] isolated a further thermostable substance, "cocarboxylase", which was necessary for the decarboxylation of pyruvate. In 1937, Lohmann and Schuster [32] investigated the constitution of "cocarboxylase" and found it to be a diphosphate ester of vitamin B1 (ThDP) **3** (Scheme 2). Three years later, Green and co-workers [33] proposed that magnesium functions as a "bridge" between the apoenzyme and ThDP. This hypothesis was supported several years later by ^{31}P-NMR [34], and has unambiguously been proved by X-ray spectroscopy [35].

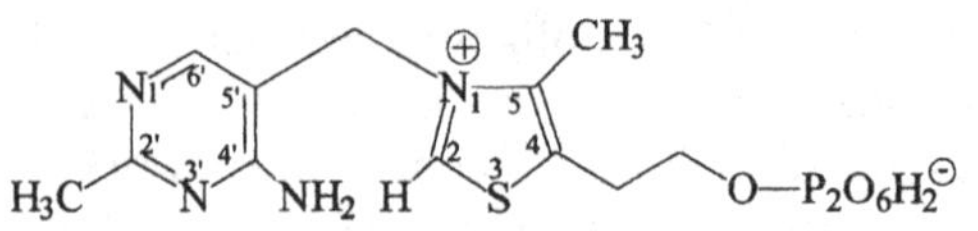

ThDP **3**

Scheme 2. Thiamine diphosphate

First steps to elucidate the reaction mechanism of PDC were achieved by the investigation of model reactions using ThDP or thiamine [36, 37]. Besides the identification of C2-ThDP as the catalytic center of the cofactor [36], the mechanism of the ThDP-catalyzed decarboxylation of α-keto acids as well as the formation of acyloins was explained by the formation of a common reaction intermediate, "active acetaldehyde". This active species was first identified as HEThDP **7** (Scheme 3) [38, 39]. Later studies revealed the α-carbanion/enamine **6** as the most likely candidate for the "active acetaldehyde" [40–47] (for a comprehensive review see [48]). The relevance of different functional groups in the ThDP-molecule for the enzymatic catalysis was elucidated by "site-directed substitutions" of the cofactor ThDP by chemical means (for a review see [49, 50]). The recent analysis of the 3D-structure based on X-ray data of PDCs [35, 51] opened a new field of research. Based on the structural data, site-directed mutagenesis studies allowed investigation of the relevance of certain amino acid side-chains for the binding of the cofactors, the catalytic mechanism, and the stability of the enzyme.

2 Approaches to the Reaction Mechanism

PDC increases the rate of decarboxylation of pyruvate by thiamine alone by a factor of 3×10^{12} at pH 6.2 and 30 °C [52]. The capacity of ThDP to catalyse the decarboxylation of α-keto acids depends mainly on two properties of the thiazolium ring of ThDP: (**a**) its capacity to ionize to form a nucleophilic anion and thus bind to the α-carbonyl group of pyruvate, and (**b**) its ability to stabilize the negative charge upon cleavage of carbon dioxide.

The different steps which are relevant for thiamine-catalyzed decarboxylation and the formation of α-hydroxy ketones are summarized in Scheme 3.

The reaction cycle is started with the activation of ThDP by the enzyme. For this common step in all ThDP-dependent enzymes, a stepwise and a concerted

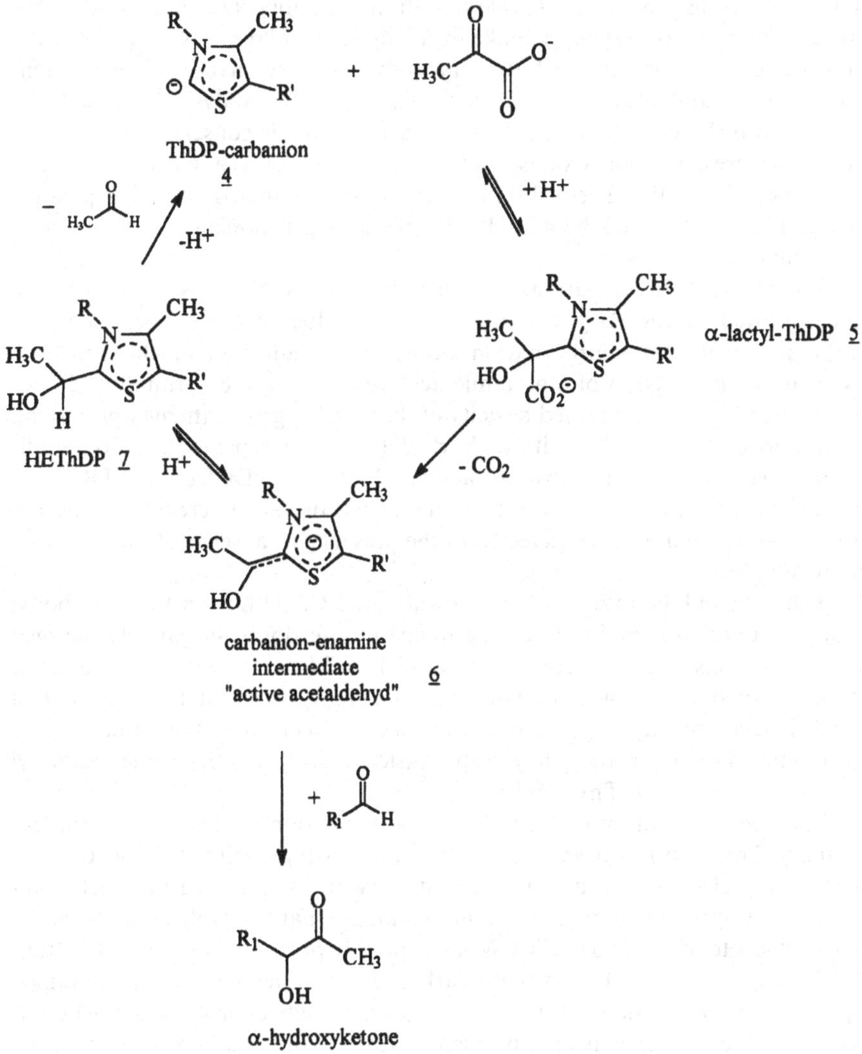

Scheme 3. Reaction path of enzymatic pyruvate decarboxylation and formation of α-hydroxy ketones

pathway are controversially discussed. The stepwise pathway assumes the fast dissociation of the C2 proton in the enzyme-bound state, which makes the stabilization of the ThDP-carbanion necessary [50, 53]. Other authors favour a concerted pathway, which assumes the deprotonation of C2 simultaneously with the addition of pyruvate [54–57]. The structural data suggested that the deprotonation of the C2 atom is catalyzed by the cofactor (4'-NH_2-group) itself, since no enzymatic group lies close enough to this proton to act as the base that initiates catalysis [35, 58, 59]. This assumption was supported by recent molecular modeling studies [60] which make the involvement of the 4'-NH$_2$-group in the protonated and deprotonated form very likely as an explanation of the different protonation and deprotonation steps during enzymatic catalysis (Scheme 3).

A carboxylic side-chain of a glutamic acid (which is conserved in pyruvate decarboxylases, pyruvate oxidase and transketolase) forms a hydrogen bond to 1'N of ThDP, and thus keeps this nitrogen in the protonated form. The positive charge is delocalized to the 4'-NH$_2$-group, making it possible for it to act as a proton donor [35, 60].

Recent NMR-investigations revealed that the active C2-site of ThDP is undissociated in the enzyme-bound state and does not exist as a discrete carbanion. The fast deprotonation occurs upon addition of the substrate pyruvate by interaction of a glutamic acid with N1' in the pyrimidine ring of ThDP, leading to an increased basicity of the 4'-NH$_2$-group, thereby producing an intermediate carbanion with a short lifetime. The deprotonation is significantly accelerated in the substrate activated state of PDCS.c.. In PDC from *Z. mobilis*, which is not activated by the substrate, no differences in the deprotonation step have been detected in the presence or absence of the substrate pyruvate [61].

The nucleophilic attack of the deprotonated C2-ThDP on the α-carbonyl group of pyruvate results in a very unfavourable double-negatively charged species. As has been deduced from model building studies, this species is stabilized by one hydrogen bond of the carboxylate group with the side-chain of Glu477. A second hydrogen bond is formed between the alkoxy ion and the protonated 4'-NH$_2$-group. This group transfers a proton to the former carbonyl group to give α-lactyl-ThDP **5** [60, 62].

The decarboxylation of **5** results in the formation of the α-carbanion/enamine **6**. This step has been intensively studied using conjugated α-keto acids with strong electron-withdrawing substituents on the phenyl ring. Such compounds are converted to an α-carbanion/enamine that is a visible chromophore with a discrete lifetime [64–65]. **6** is subsequently protonated to give HEThDP **7**. This step is reversible, and the α-carbanion/enamine may also be obtained upon binding of acetaldehyde to ThDP and subsequent deprotonation. Alternatively a concerted mechanism is discussed, which avoids **6** as a discrete enzyme-bound intermediate [55]. The latter was demonstrated during acetoin formation from acetaldehyde by PDCS.c.. The cleavage of acetaldehyde from HEThDP **7**, followed by deprotonation results in the regeneration of the ThDP-carbanion **4**.

3 The 3D-Structure

The 3D-structure of the α$_4$ isoenzyme of PDC from *S. uvarum* (PDCS.u.) was reported by Dyda and co-workers in 1993 [35], a short time after the publication of the crystal structures of two other ThDP-dependent enzymes, the transketolase from *S. cerevisiae* [66] and the pyruvate oxidase from *L. plantarum* [67, 68]. Recently Arjunan et al. (1996) published the 3D-structure of the α_4 isoenzyme of PDC from *S. cerevisiae* (PDC*S.c.*) [51]. Both PDC structures proved to be very similar. A detailed description and comparison of the structures of ThDP-dependent enzymes is given in Ref. [58, 59].

The polypeptide chain of PDC is organized in three domains. The amino-terminal domain (α) binds the pyrimidine ring, whereas the carboxy-terminal domain (γ) forms the diphosphate binding site, while the middle domain (β) performs the intermolecular contact between the two dimeric halves in the tetramer and contains the substrate regulation site in PDC*S.c.* (details about the substrate regulation behaviour in PDC*S.c.* are discussed in Sects. 4 and 6.1).

Two active centers are formed at the interface of two subunits, building up one dimer such that the α-domain of one subunit interacts with the γ-domain of the second subunit and vice versa. The interface between the two dimeric halves of the tetrameric molecule is not directly involved in the catalytic center.

Since there are few side-chain contacts between the dimers compared with those stabilizing one single dimer, the tetrameric structure of PDC may be described as a dimeric dimer.

Those amino acids which are directly or indirectly involved in ThDP binding are conserved in transketolase, pyruvate oxidase and PDC. Amongst these are the residues which form the metal-binding site and most significantly, a conserved glutamic acid residue which forms a hydrogen bond to the N1'-atom of the pyrimidine ring (see Sect. 2). The cofactor is buried in a mostly hydrophobic pocket, such that only the S, C2 and N4'-atoms are accessible from the solvent. In all three enzymes ThDP is bound in the "V"-form (Scheme 2). This conformation brings the 4'NH$_2$-group within interaction distance to the C2 reaction center.

The amino acid residues which are involved in the binding of the cofactors (Table 1) and those which are located in the cavity leading to the active center (Table 2) are very similar among PDCs from various organisms (Figs. 1 and 2).

3.1 Stability of the Tetrameric Enzyme

3.1.1 Apo-PDC and Holo-PDC

The reversibility of the cofactor binding by PDC was demonstrated in 1941 by Green and co-workers [33] and by Kubowitz and Lüttgens [69] (Scheme 4). Schellenberger confirmed these results in 1967 and postulated a first mechanistical model [40].

$$\text{apo-PDC} + 4\ \text{ThDP} + 4\ \text{Mg}^{2+} \underset{> \text{pH } 8{,}0}{\overset{\text{pH } 6{,}5}{\rightleftharpoons}} \text{holo-PDC}$$

Scheme 4.

Table 1. Comparison of amino acid residues which are involved in the binding of the cofactors ThDP and Mg^{2+} in PDCS.c. and PDCZ.m., respectively. The corresponding amino acid residues from PDCZ.m. have been deduced from sequence alignments (Figs. 1 and 2) based on the X-ray structure of PDCS.u. [35] and PDCS.c. [51]. Different residues are underlined. Those which have been exchanged by site-directed mutagenesis are marked with * (see Tables 4 and 5)

Cofactor	PDCS.c.	PDCZ.m.
Mg^{2+}	Asn 444	Asn 440*
	Asn 471	Asn 467*
	Gly 473	Gly 469
ThDP (diphosphate)	Thr 390	Asp 390
	Ser 446	Ser 442
	Ile 476	Ile 472
ThDP (thiazolium ring)	Ile 415*	Ile 415
ThDP (aminopyridine ring)	Glu 51*	Glu 50*
	Gly 413	Gly 413
	Thr 73	Thr 72
	Val 76	Val 75
	Asn 83	Asp 82
	Leu 25	Val 24
	Leu 449	Leu 445

Table 2. Comparison of amino acid residues which are involved in the cavity leading to the reaction center in PDCS.c. and PDCZ.m., respectively. The corresponding amino acid residues from PDCZ.m. have been deduced from sequence alignments (Figs. 1 and 2) based on the X-ray structure of PDCS.u. [35] and PDCS.c. [51]. Different residues are underlined. Those which have been exchanged by site-directed mutagenesis are marked with * (see Tables 4 and 5)

	PDCS.c.	PDCZ.m.
Hydrophobic residues towards the thiazolium-ring side	Ile 480	Ile 476
	Ile 476	Ile 472
	Phe 393	Phe 393
	Val 410	Val 410
	Thr 388	Thr 388
	Ala 392	Trp 392*
Hydrophilic residues towards the pyrimidin-ring side	Asp 28*	Asp 27
	Glu 477*	Glu 473
	Thr 388	Thr 388
	His 114	His 113*
	His 115	His 114

Later studies revealed that ThDP is nearly completely removable under slightly alkaline conditions, whereas it is difficult if not impossible to remove Mg^{2+} completely from the enzyme [70–74].

This dissociation equilibrium is even present at neutral pH, but to a much lower extent. As a consequence, small amounts of cofactors, especially ThDP, have to be added to the enzyme solution in order to guarantee PDC stability [70, 75, 76].

Mg^{2+} can be replaced by other divalent ions, such as Mn^{2+}, Ni^{2+}, Co^{2+}, Ca^{2+} [40], without affecting the maximal velocity (V_{max}), but the stability of the reconstructed holoenzyme is sensitive to the nature of the divalent cation [53, 77]. During their studies on the recombination of apo-PDC to holo-PDC*S.c.*, Schellenberger and co-workers [40, 49, 50, 53, 57] elucidated the significance of the different functional groups of the ThDP-molecule by incorporating various ThDP-derivatives into PDC. These "site-directed substitutions" demonstrated that only those ThDP-derivatives retaining a complete N-1'-4'-amino system proved to function as active co-enzymes with full binding capacity [49, 50]. A further significant result of these kinetic studies established that Mg^{2+} combines firmly (quasi-irreversibly) with the apoenzyme only in the presence of ThDP and vice versa, while both cofactors bind to the apoenzyme independently of each other [40]. These results have been supported by Vaccaro et al. [74], and have also been demonstrated for PDC*Z.m.* [72].

From gelfiltration studies, a molecular mass of 240 kDa was calculated for the tetrameric holoenzyme of PDC*S.c.*, while the apoenzyme was found to be a dimer at pH 8.5 [78, 79]. The pH-dependent dissociation/association of the tetramer was further investigated by X-ray solution scattering [80]. Native PDC from yeast is completely tetrameric in the range of 5.5 < pH < 6.5, between pH 7.5 and 9.5 there is an equilibrium between the dimers and the tetramers and above pH 9.5 the solutions exclusively contain dimers. The situation is different for PDC*Z.m.*, where only tetramers are found, independent of the pH-value [72, 81, 82]. CD-investigations revealed that the alkali-induced dissociation of tetrameric holo-PDC*S.c.* occurs in two steps, where the dissociation of cofactors is fast, and the dissociation of the tetramer into dimers is rate-limiting [83].

The influence of the effectors pyruvamide and phosphate on the quarternary structure of PDC*S.c.* has been demonstrated by X-ray solution scattering [113]. Whereas phosphate stabilizes the tetrameric structure, the addition of pyruvamide gave rise to a less compact tetramer, due to structural changes upon binding to the regulatory site (for a more detailed description of the substrate activation behaviour see Sects. 4.2 and 6.1).

3.1.2 Reversible Dissociation and Unfolding

The reversible dissociation and unfolding of the tetrameric enzyme mirrors the differences in protein stability between PDC*S.c.* [84] and PDC*Z.m.* [82]. The stepwise unfolding of PDC*Z.m.* using urea or guanidine hydrochloride resulted in a three-step process where the deactivation as a consequence of cofactor dissociation (1–2 M urea) precedes the dissociation of the tetramer into monomers (>3 M urea) followed by unfolding of the protein chains (>5 M urea) [82]. During the whole process the tetramer is the main species, and dimers only appear in small amounts [82]. The situation is different with PDC*S.c.*, where a structured inactive intermediate (problably dimeric) has been observed in the range of 1–2 M urea. The complete unfolding of the protein chains from PDC*S.c.* has been observed > 4 M urea [84]. From these data, a significantly weaker dimer–dimer interaction can be deduced for PDCS.c. in comparison to PDCZ.m.. Since the dissociation of PDCZ.m. into monomers results in one step, the degree of binding between the monomers is likely to be equivalent to the binding strength between the dimers forming the tetramer. By contrast, in PDCS.c. the stability of the dimer predominates that of the tetramer.

4 Pyruvate Decarboxylases from Various Species

PDC (E.C. 4.1.1.1) is the first enzyme of the branched glycolytic pathway that together with alcohol dehydrogenase (ADH) (E.C. 1.1.1.1) converts pyruvate to ethanol under anaerobic conditions. During aerobiosis, pyruvate is converted to acetyl-CoA by pyruvate dehydrogenase; during anaerobiosis, lactate or ethanol is formed to oxidize the NADH. The pathway, which produces ATP by substrate-level phosphorylation, is important for the survival of plants that experience hypoxic or anoxic conditions that occur when plants are submerged during flooding [85].

Both fermentation and respiration may contribute to glucose catabolism in the course of the aerobic growth of yeast. In *Saccharomyces cerevisiae*, the respiratory and fermentive metabolisms are tightly regulated and mutually exclusive. The regulatory mechanism of glucose repression allows the final product of glycolysis, pyruvate, to be almost completely channelled by PDC to acetaldehyde, and then to ethanol by the action of ADH. The majority of other yeast species however, rely on a predominantly respiratory metabolism of glucose under aerobic conditions [86].

Genes for pyruvate decarboxylases have been isolated from yeasts and fungi such as *Saccharomyces cerevisiae* [86–90], *Hanseniaspora uvarum* [91], *Klyveromyces marxianus* [92], *Klyveromyces lactis* [93], *Neurospora crassa* [94], *Aspergillus parasiticus* [95], plants, like maize (*Zea maize*) [96, 97], rice (*Oryza sativa*) [98], tomato (*Lycopersicon esculentum*) [99, 100], pea (*Pisum sativum*) [101], and tobacco (*Nicotiana tabacum*) [102] and from the bacterium *Zymomonas mobilis* [103–106].

All these PDC sequences consist of subunits of about 562–610 amino acid residues (Table 3a and b, Figs. 1 and 2). Enzymes from plants (pea, maize, rice) have somewhat longer chains compared to the enzymes from yeasts, fungi and bacteria. The sequence alignment in Fig. 2 demonstrates that the elongation of the plant-PDC sequences concerns the N-terminus.

Despite the large number of available PDC-genes, only a few enzymes have been characterized in detail with regard to their protein stability, relevance of functional groups and kinetic properties. The enzymes from *Saccharomyces cerevisiae* [34, 40–47, 62–64, 74, 75, 77–80, 83, 107–116] and *Zymomonas mobilis* [72, 76, 82, 106, 117–119], and *Pisum sativum* [125, 126] have been characterized in detail. Some data are available for the enzymes from wheat germ

[120–124,133], whereas the pyruvate decarboxylases from *Zea maize* (from kernels and roots) [71,126,127], *Oryza sativa* [128], *Hanseniaspora uvarum* [91], sweet potato (*Ipomoae batatas L.*) [129,130], lima bean [126] and *Sarcina ventriculi* [131] have only partially been characterized.

4.1 Active Association States

For all pyruvate decarboxylases the tetramer is the active species (Table 3a and b). Higher active association states have been described for PDC from maize [127], pea [125], and wheat germ [124]. For PDC*Z.m.*, small amounts of active octamers have been detected, depending on the protein concentration [81,82].

The loss of cofactors is accompanied by a decrease in the degree of oligomerization for PDC*S.c.* [80] and peas [125] (see Sect. 3.1.1).

Table 3a. Pyruvate decarboxylases of various yeasts, fungi and bacteria

Species	amino acids/ subunit (kDa)	$S_{0.5}/V_{max}$ (mM)/(U/mg)	Hill-coeff. n_h	Ref.
Yeasts/Fungi				
Saccharomyces cerevisae	562 (PDC1) 61.468 [1]	α_4: 1.16/62 β_4: 1.14/65	α_4 1.93, 2.33 β_4 2.02	[83, 87–90]
	563 (PDC5) 62.0 [1]	$(\alpha\beta)_2$: n.d.	$(\alpha\beta)_2$ 2.11	[108, 114, 123]
	α 55–59 [2] β 59–63 [2]	mixed PDC [3] 1.0/45–60	mixed PDC [3] 1.8	
Hanseniaspora uvarum	564 61.070 [1] 57 [2]	0.75/53	n.d.	[91]
Klyveromyces marxianus	564 61.9 [1]	n.d./n.d.	n.d.	[92]
Klyveromyces lactis	563	n.d./n.d.	n.d.	[93]
Neurospora crassa	570 62.263 [1]	n.d./n.d.	n.d.	[94]
Aspergillus parasiticus	563 64.0	n.d./n.d.	n.d.	[95]
Bacteria				
Zymomonas mobilis	568 60.925 [1]	0.4–1.1/ 120–180	1.0	[103–106] [117–119]
Sarcina ventriculi	n.d.	13.0/103	> 1.0	[131]

[1] Molecular mass determined from amino acid composition deduced from DNA sequence
[2] Molecular mass determined from SDS-PAGE or gelfiltration.
[3] Mixed PDC from yeast contains a mixture of 3 isoenzymes α_4, $(\alpha\beta)_2$, β_4.
n.d. = no data

Table 3b. Pyruvate decarboxylases of various plants

Species	amino acids/ subunit (kDa)	$S_{0.5}/V_{max}$ (mM)/(U/mg)	Hill-coeff. n_h	Ref.
Monocots				
Zea maize (maize)	610 65.427 [1]	1.0/96	2.5/3.2 [3] 1.4/1.8 [4]	[71, 96, 97] [127]
Oryza sativa (rice)	602 65.066 [1] α 62.0 [2] β 64.0 [2]	0.25/60	1.65	[98, 128]
Triticum aestivum (wheat)	α 61.5 [2] β 64.0 [2] 390 (nativ)	3.0/51	1.5 1.9	[122–124] [133]
Dicots				
Pisum sativum (pea)	593 63.963 [1] α 65 [2] β 68 [2]	1.0/45	>1	[101, 125] [126]
Nicotiana tabacum (tobacco)	614 67.1 [1]	n.d.	n.d.	[102]
Ipomea batatas (sweet potato)	60.0 [2]	0.6/12.3	> 1.0	[129, 130]
Lycopersicon esculentum (tomato)	α 62.0 [2] β 64.0 [2]	n.d./n.d.	n.d.	[99, 100]

[1] Molecular mass determined from amino acid composition deduced from DNA sequence.
[2] Molecular mass determined from SDS-PAGE or gelfiltration.
[3] in kernels (pH-dependent)
[4] in roots (pH-dependent)
n.d. = no data

4.2 Kinetic Properties

With the exception of PDCZ.m., all enzymes which have been investigated with respect to their kinetical data (Table 3a and b) exhibit sigmoidal v/S-plots (Hill coefficient > 1), and a lag-phase during product formation. Both effects point to substrate activation behaviour, and either pyruvate or the substrate surrogate pyruvamide function as an activator. The pyruvate concentration needed to obtain half-maximal velocity ($S_{0.5}$) is between 0.25 mM and 1.1 mM pyruvate. The enzyme from *Sarcina ventriculi* shows the comparatively highest $S_{0.5}$-value at 13 mM. The maximal velocities (V_{max}) are in the range of 30–180 U/mg. The enzymes from yeasts and plants exhibit specific activity in the range of 45–60 U/mg. By contrast, the bacterial enzymes from *Z. mobilis* and *S. ventriculi* are 2 or 3 times more active. The pH-optima are in the range of 6.0–6.8. Phosphate proved to be a competitive inhibitor for the enzymes from *S. cerevisiae* [132], potato [130], maize [127], rice [128], pea [125] and wheat germ [133], but not for the bacterial enzymes [76, 82, 131].

```
  1 ------SEITLGKYLFERLKQVNVNTVFGLPGDFNLSLLDKIYEVEGMRWAGNANELNAR Scpdc1
  1 -----MSEITLGKYLFERLSQVNCNTVFGLPGDFNLSLLNKLYEVKGMRWAGNANELNAA Scpdc5
  1 -----MSEITLGRYVFERIKQVGVNTIFGLPGDFNLSLLDKIYEVEGLRWAASLNELNAA Hupdc
  1 -----MSEITLGRYLFERLKQVEVQTIFGLPGDFNLSLLDKIYEVPGMRWAGNANELNAA Kmpdc
  1 MVAQQQGKFTVGDYLAERLAQVGVRHHFVVPGDYNLILLDKLQAHPDLKEVGCANELNCS Ncpdc


 55 YAADGYARIKGMSCIITTFGVGELSALNGIAGSYAEEVGVLHVVGVPSISSQAKQLLLHH Scpdc1
 56 YAADGYARIKGMSCIITTFGVGELSALNGIAGSYAEEVGVLHVVGVPSISSQAKQLLLHH Scpdc5
 56 YAADGYSRIKGLGVIITTFGVGELSALNGIAGAYAEEVGVLHIVGVPSLASQAKQLLLHH Hupdc
 56 YAADGYARLKGMACVITTFGVGELSALNGIAGSYAEEVGVLHVVGVPSISSQAKQLLLHH Kmpdc
 61 LAAEGYARANGISACVVTYSVGALSAFNGTGSAYAENLPLVLISGSPNTNDPSQYHILHH Ncpdc


115 TLGNGDFTVFHRMSANISETTAMITDICTAPAEIDRCIRTTYVTQRPVYLGLPANLVDLN Scpdc1
116 TLGNGDFTVFHRMSANISETTAMITDIRNAPAEIDRCIRTTYTTQRPVYLGLPANLVDLN Scpdc5
116 TLGNGDFDVFHRMSANISETTAMITDLAAAPAEIDRCIRTAYIAQRPVYLGLPANLVDLN Hupdc
116 TLGNGDFTVFHRMSSNISETTAMITDINSAPSEIDRCIRTTYISQRPVYLGLPANLVDLK Kmpdc
121 TLGHPDYTYQYEMAKKITCCAVAIPRAIDAPRLIDRALRAAILARKPCYIEIPTNLAGAT Ncpdc


175 VPAKLLQTPIDMSLKPNDAESEKEVIDTILVLAKDAKNPVILADACSRHDVKAETKKLI Scpdc1
176 VPAKLLETPIDLSLKPNDAEAEAEVVRTVVEFIKDAKNPVILADACSRHDVKAETKKLM Scpdc5
176 VPAKLLETKIDLALKANDAEAENEVVETILALVADAKNPVILSDACASRHNVKAEVKQLI Hupdc
176 VPASLLETPIDLSLKPNDPEAENEVLETVLELIKDAKNPVILADACSRHNVKAETKKLI Kmpdc
181 CVRPGPISAITDPITSDKSALEAAA-KCAAEYLDGKLKPVILVGPKAGRAGSEKELIEFA Ncpdc


235 DLTQFPAFVTPMGKGSISEQHPRYGGVYVGTLSKPEVKEAVESADLILSVGALLSDFNTG Scpdc1
236 DLTQFPVYVTPMGKGAIDEQHPRYGGVYVGTLSRPEVKKAVESADLILSIGALLSDFNTG Scpdc5
236 DATQFPAFVTPLGKGSIDEKHPRFGGVYVGTLSSPEVKQSVESADLILSVGALLSDFNTG Hupdc
236 DITQFPAFVTPMGKGSIDEQHPRFGGVYVGTLSSPEVKEAVESADLVLSVGALLSDFNTG Kmpdc
240 EAMGCAVALQPAAKGMFPEDHKQFVGIFWGQVSSDAADAMVHWADAMICVGAVFNDYSTV Ncpdc


295 SFSYSYKTKNIVEFHSDHMKIRNATFPGVQMKFVLQKLLTNIADAAKGYKPVAVPARTPA Scpdc1
296 SFSYSYKTKNIVEFHSDHIKIRNASFPGVQMKFALQKLLDAIPEVAKDYKPVAVPARVPI Scpdc5
296 SFSYSYQTKNIVEFHSDYIKIKNASFPGVQMKFVLEKLIAKVGAKIANYSPVPVPAGLPK Hupdc
296 SFSYSYKTKNIVEFHSDYIKVRNATFPGVQMKFVLQKLLTKVKDAAKGYKPVPVPHAPRD Kmpdc
300 GWTAVPNIPLMTVDMDHVTFPGAHFSRVRMCEFLSHLATQVTFNDSTMIEYKRLKPDPPH Ncpdc


355 NAAVPASTPLKQEWMWNQLGNFLQEGDVVIAETGTSAFGINQTTFPNNTYGISQVLWGSI Scpdc1
356 TKSTPANTPMKQEWMWNQLGNFLREGDIVIAETGTSAFGINQTTFPTDVYAIVQVLWGSI Scpdc5
356 NAPVADSTPLAQEWLWNELGEFLEEGDIVVTETGTSAFGINQTRFPTDAYGISQVLWGSI Hupdc
356 NKPVADSTPLKQEWVWTQVGKFLQEGDVVLTETGTSAFGINQTHFPNDTYGISQVLWGSI Kmpdc
360 VHTAEREEPLSRKEISRQVQEMLTDKTSLFVDTGDSWFNGIQLKLPPGAKFEIEMQWGHI Ncpdc


415 GFTTGATLGAAFAAEEIDPKKRVILFIGDGSLQLTVQEISTMIRWGLKPYLFVLNNDGYT Scpdc1
416 GFTVGALLGATMAAEELDPKKRVILFIGDGSLQLTVQEISTMIRWGLKPYIFVLNNNGYT Scpdc5
416 GYSVGAMVGATFAAEELDKAKRVILFVGDGSLQLTVQEIACLIRWGLKPYIFVLNNGYT Hupdc
416 GFTGGATLGAAFAAEEIDPKKRVILFIGDGSLQLTVQEISTMIRWGLKPYLFVLNNDGYT Kmpdc
420 GWS----IPAAFGYALRHPDRHTIVLVGDGSFQVTAQEVSQMVRFKVPITIMLINNRGYT Ncpdc


475 IEKLIHGPKAQYNEIQGWDHLSLLPTFGAKDYETHRVATTGEWDKLTQDKSFNDNSKIRM Scpdc1
476 IEKLIHGPHAEYNEIQGWDHLALLPTFGARNYETHRVATTGEWEKLTQDKDFQDNSKIRM Scpdc5
476 IEKLIHGPTAQYNMIQNWKQLRYLTNFGATDYEAIPVKTVGEWKKLTADPAFKKNSTIRL Hupdc
476 IERLIHGETAQYNCIQSWKHLDLLPTFGAKDYEAVRVATTGEWNKLTTDKKFQENSKIRL Kmpdc
476 IEVEIHDGSYNKIKNWDYAMLVEAFNSTDGHAKGLLANTAGELADAIKVAESHKEGPT-L Ncpdc


535 IEVMLPVFDAPQNLVEQAKLTAATNAKQ-------- Scpdc1
536 IEVMLPVFDAPQNLVKQAQLTAATNAKQ*------- Scpdc5
536 IEVFLPEMDAPSSLVAQANLTAAINAKQD------- Hupdc
536 IEVMLPVMDAPSNLVKQAQLTASINAKQE------- Kmpdc
535 IECTIDQDDCSKELITWGHYVAAANARPPRNMSVQE Ncpdc
```

Fig. 1. Sequence alignment of different pyruvate decarboxylases from yeasts and fungi. The alignment was performed using the program HUSAR from EMBL (Heidelberg, Germany). The ThDP-binding motiv [173] is underlined. Catalytically important residues are shaded (see text). **Scpdc1:** PDC1 from *S. cerevisiae*; **Scpdc5:** PDC5 from *S. cerevisiae*; **Hupdc:** PDC from *Hanseniaspora uvarum*; **Kmpdc:** PDC from *Klyveromyces marxianus*, **Ncpdc:** PDC from *Neurospora crassa*

```
  1 --------------------------------------------MSYTVGTYLAERLVQ Zympdc
  1 METLLAGNPANGVAKPTCNGVGALPVANSHAIIATPAAAAATLAPAGATLGRHLARRLVQ Zmpdc1
  1 MELALVGNPSNGVAKPSCNSVGSLPVVSSNAVIHPPVTSAAR-----ATLGRHLARRLVQ Orspdc1
  1 -----------------METETETPNGSTPCPTSAPSAIPLRPSSCDGTMGRHLARRLVE Pspdc1
  1 --------------------------------------------MEGETLPLAQYLFKRLLQ Asppdc

 16 IGLKHHFAVAGDYNLVLLDNLLLNKNMEQVYCCNELNCGFSAEGYARAKGAAAAVVTYSV Zympdc
 61 IGASDVFAVPGDFNLTLLDYLIAEPGLTLVGCCNELNAGYAADGYARSRGVGACAVTFTV Zmpdc1
 56 IGATDVFAVPGDFNLTLLDYLIAEPGLKLIACCNELNAGYAADGYARA--LVGAFVTFTV Orspdc1
  1 ------------------------------AADGYARARGVGACVVTFTV TbPDC1
 44 IGVRDVFSVPGDFNLTLLDHLIAEPELNLVGCCNELNAGYAADGYGRAKGVGACVVTFTV Pspdc1
 19 LGVDSIFGVPGDYNLTLLDHVVP-SGLKWVGNCNELNAGYAADGYSRIKDIGAVVTTFGV Asppdc

 76 GALSAFDAIGGAYAENLPVILISGAPNNNDHAAGHVLHHALGKTDYHYQLEMAKNITAAA Zympdc
121 GGLSVLNAIAGAYSENLPVVCIVGGPNSNDYGTNRILHHTIGLPDFSQELRCFQTITCYQ Zmpdc1
114 GGLSVLNAIAGAYSENLPVICIVGGPNSNDYGTNRILHHTIGLPDFSQELRCFQTITCYQ Orspdc1
 21 GGLSVLNAIAGAYSENLPLICIVGGPNSNDYGTNRILHHTIGLQDFSQEPRCFQTVTCYR TbPDC1
104 GGLSILNAIAGAYSENLPVICIVGGPNSNDYGTNRILHHTIGLPDFSQELQCFQTITCFQ Pspdc1
 78 GELSAINAIAGAYAEKAPVVHIVGTPMRASQESRALIHHTFNDGDYQRFDAIQEHVTVAQ Asppdc

136 EAIYTPEEAPAKIDHVIKTALREKKPVYLEIACNIASM--PCAAPGPASALFNDEASDEA Zympdc
181 AIINNLDDAHEQIDTAITATALRESKPVYISVSCNLAGLSHPTFSRDPVPMFISPRLSNKA Zmpdc1
174 AVINNLDDAHEQIDTAITATALRESKPVYISVGCNLAGLSHPTFSREPVPLFISPRLSNKA Orspdc1
 81 AVVNNLEDAHELIDTAVSTALKESKPVYISIGCNLPGIPHPTFSREPVPFFALSPRLSNMM TbPDC1
164 AVVNNLDDAHELIDTAISTALKESKPVYISIGCNLPAIPHPTFARDPVPFFLAPRVSNQA Pspdc1
138 VSLSDHRTAPSEIDRILLQCLLHSRPVRIAIPVDMVPVLVPVAGLSSKIQ-IPPAVRQPQ Asppdc

194 SLNAAVEETLKFIANRDKVAVLVGSELRAAGAEEAAVKFTDALGGAVATMAAAKSFFPEA Zympdc
241 NLEYAVEAAADFLNKAVKPVMVGGPKIRVAKAREAFAAVADASGYPFAVMPAAKGLVPEH Zmpdc1
234 NLEYAVEAAADFLNKAVKPVMVGGPKIRVAKAKKAF-AIAESSGYPFAVMPSAKGLVPEH Orspdc1
141 GLEAAVEAAAEFLNKAVKPVLVGGPKMRVAKASDAFVELSDACGYAVAVMPSAKGLFPEH TbPDC1
224 GLEAAVEEAAAFLNKAVKPVIVGGPKLRVAKAQKAFMEFAEASGYPIAVMPSGKGLVPEN Pspdc1
197 AEEAALNAVLKRIYSSKKPMILVDGITRSFGMLQRVNHFIQTIGWPTFTSGFGKGLVDET Asppdc

254 NPHYIGTSWGEVSYPGVEKTMKEADAVIALAPVFNDYSTTGWTDIPDPKKLVLAEPRSVV Zympdc
301 HPRFIGTYWGAVSTTFCAEIVESADAYLFAGPIFNDYSSVGYSLLLKREKAVIVQPDRMV Zmpdc1
293 HPRFIGTYWGAVSTTFCAEIVESADAYLFAGPIFNDYSSVGYSLLLLKREKAVIVQPDRVV Orspdc1
201 HSHFIGTYWGAVSTAFCAEIVESADAYLFAGPIFNDYSSVGYSLLLKKEKAIIVQPDRVT TbPDC1
284 HPHFIGTYWGAVSTSYCGEIVESADAYVFVGPIFNDYSSVGYSLLIKKEKSLIVQPNRVT Pspdc1
257 LPNVYG-----VCTLHQKAFVDSCDLVLVFGPHFSNTNSYNYFLKPADEKSVLFSPNSIQ Asppdc

314 VRRIRFPSVHLKDYLTRLAQKVSKKTGALDFFKSLNAGELKKAAPADPSAPLVNAEIARQ Zympdc
361 VGDGPAFGCILMPEFLRALAKRLRRNTTAYDNYRRIFVPDREPPNGKPNEPLRVNVLFKH Zmpdc1
353 VGNGPAFGCILMTEFLDALAKRLDRNTTAYDNYRRIFIPDREPPNGQPDEPLRVNILFKH Orspdc1
261 IGNGPAFGCVLMRDFLAALAKRLKHNPTAFENYHRIYVPEGHPLKCEPKEALRVNVLFQH TbPDC1
344 IGNGLSLGWVFMADFLTALAKKVKTNTTAVENYRRIYVPPGIPLKREKDEPLRVNVLFKH Pspdc1
312 VNKDVFRDLPVGYFIEQLTQQLDISRIPTHKHDLVHPSLRTLPEVSPTDLVTQTGGFWKR Asppdc

374 VEALLTPNTTVIAETGDSWFNAQRMKLPNGARVEYEMQWGHIGWSVPAAFGYAVGAPERR Zympdc
421 IKGMLSGDSAVVAETGDSWFNCQKLRLPEGCGYEFQMQYGSIGWSVGATLGYAQAAKD-- Zmpdc1
413 IKEMLSGDTAVIAETGDSWFNCQKLRLPEGCGYEFQMQYGSIGWSVGATLGYAQAAKD-- Orspdc1
321 IQNMLSGDSVVIAETGDSWFNCQKLKLPKGCGYEFQMQYGSIGWSVGATLGYA------- TbPDC1
404 IQALISGDTAVIAETGDSWFNCQKLRLPENCGYEFQMQYGSIGWSVGATLGYAQAATD-- Pspdc1
372 FSPFLRTGDIILGETGTPGYGVNDFILPPQTRLFKPATWLSIGYMLPAALGASHAQRDLV Asppdc

434 NILM------------VGDGSFQLTAQEVAQMVRLKLPVIIFLINYGYTIEVMI--HDG Zympdc
479 ----------KRVIACIGDGSFQVTAQDVSTMLRCGQKSIIFLINNGGYTIE--VEIHDG Zmpdc1
471 ----------KRVISCIGDGSFQMTAQDVSTMLRCGQKSIIFLINNGGYTIE--VEIHDG Orspdc1
374 -----QAAPEKRVIACIGDGSFQVTAQDISTMLRCGQRTIIFLINNGGY----------- TbPDC1
462 ----------KRVIACIGDGSFQVTAQDISTMLRCGQRSIIFLINNGGYTIE--VEIHDG Pspdc1
432 ASDQYHSLSNPRTILFIGDGSFQMTVQELSTIIHQKLNVIIFLINNDGYTIERCIHGRNQ Asppdc

480 PYNNIKNWDYAGLMEVFNGNGGYDSGAAKGLKAKTGGELAEAIKVALANTDGPTLIECFI Zympdc
527 PYNVIKNWDY---TGLVNAIHNSEGNCWTMKVRTEEQLKEAIATVTGAKKDCLCFIEVIV Zmpdc1
519 PYNVIKNWDY---TGLIDAIHNSDGNCWTKKVRTEEELIEAIATATGAKKDCLCFIEIIV Orspdc1
510 PYNVIKNWDY---TGFVSAIHNGQGKCWTAKVRTEEDLTEAIATATGAEKDSLCFIEVFA Pspdc1
492 AYNDVAPWRYLKAAEFFGADQDGEYKASTWEVRTWADLDRVLNDSQLADGKGLRMVEVFM Asppdc

540 GREDCTEELVKWGKRVAAANSRKPVNKLL* Zympdc
584 HKDDTSKELLEWGSRVSAANSRPP--NP-Q Zmpdc1
576 HKDDTSKELLEWGSRVSAANSRPP--NP-Q Orspdc1
567 HKDDTSKELLEWGSRVAAANSRPP--NPQ* Pspdc1
552 ERLDAPDVLMGLLNNQVLRENAQS--RL*- Asppdc
```

4.3 Substrate Regulation of PDCS.c.

The substrate regulation behaviour of PDCS.c. has been concluded from the sigmoidicity of the plots of velocity vs substrate concentration (v/S-plot) and a lag-phase during product formation [44, 132, 134]. Boiteux and Hess (1970) discovered the negative influence of phosphate on the enzyme affinity for pyruvate [132]. Lehmann et al. (1973) described the application of the substrate surrogate pyruvamide as an allosteric activator [115]. Some years later, Hübner and co-workers (1978) demonstrated that pyruvamide is not a competitive inhibitor of pyruvate and thus must bind to a different binding site [135].

First hints on the nature of regulatory sites in yeast PDC have been obtained from inhibition experiments with the thiol-modifying reagents 4-hydroxy-mercuri-benzoate and 3-bromo-pyruvamide in the presence and absence of the substrate pyruvate. Upon thiol modification, PDC lost its activation behaviour. Pyruvate was able to protect the enzyme against thiol-modification, suggesting that both modifier and substrate compete for the same sites on the enzyme [136]. Zeng et al. (1991) confirmed the existence of a cysteine residue near the active center [64].

The investigation of a PDC1-PDC6 fusion enzyme, retaining only Cys 221 [137], and site-directed mutagenesis studies [138] revealed Cys221 as the target of substrate activation by pyruvamide or pyruvate. The activator is likely to build a hemithioketal adduct to the sulfur of Cys221, and this adduct may easily bridge the gap between the α- and γ-domain. As a consequence, residues in both domains must be dislocated by this adduct formation [35, 51] (for a more detailed discussion of the substrate activation mechanism see Sect. 6.1). Based on these results, a model for the regulation of PDC has been suggested, in which the opening and closing of sequestering structures at the active site are driven by sulfhydryl addition/elimination reactions at the carbonyl group of the regulatory pyruvate molecule [107, 117].

4.4 Isoenzymes

Different structural genes have been reported for PDC from *Saccharomyces cerevisiae* [87–90], *Pisum sativum* [101], *Zea maize* [96, 97] and *Nicotiana tabacum* [Genbank entries X81854, X81855]. The relevance of the different structural genes in *S. cerevisiae* has been investigated in detail. PDC1 codes for the major isoenzyme and is responsible for most of the wt-PDC activity

----

Fig. 2. Sequence alignment from different pyruvate decarboxylases from *Z. mobilis* (**Zympdc**), *Zea maize* (**Zympdc1**), *Oryza sativa* (**Orspdc1**), *Nicotiana tabacum* (**TbPDC1**), *Pisum sativum* (**Pspdc1**) and *Aspergillus parasiticus* (**Asppdc**). The alignment was performed using the program HUSAR from EMBL (Heidelberg, Germany). The ThDP-binding motiv [173] is underlined. Catalytically important residues are shaded (see text)

[87–90, 139, 140]. The expression of either PDC1 (62 kDa) [140, 141] or PDC5 [90] in *E. coli* resulted in the formation of active homotetramers. It is still a point of discussion, whether the existence of different structural genes is correlated with the observation of various isoforms which have been isolated on the protein level [83, 108, 114, 123]. Alternatively the partial proteolytic digestion of one single gene product is discussed as the origin of the different isoforms with slightly different subunit sizes. This assumption has been supported by the isolation of active homotetrameric PDC from a protease-deficient *S. cerevisiae* strain, which was composed of only the larger subunit [142]. Farrenkopf and co-workers (1992) were able to separate three different isoenzymes (α_4, $(\alpha\beta)_2$, β_4) by anion exchange chromatography. The comparison of the kinetic data of the decarboxylation reaction revealed similar data for the α_4 and β_4 isoenzymes [108] (Table 3). The two isoenzymes α_4 and $(\alpha\beta)_2$ are kinetically and mechanistically distinguishable only with respect to the formation of acetoin from acetaldehyde [55].

4.5 PDCZ.m. is Unique among Pyruvate Decarboxylases

A comparison of the enzymological, kinetic, and stability data of pyruvate decarboxylases from various sources makes the special position of PDC from *Z. mobilis* obvious (Table 3a and b). Compared with other enzymes either from yeasts or plants, PDCZ.m. has only one structural gene, giving rise to the expression of a well-defined homotetramer of high stability and the highest velocity [106, 118]. With respect to the kinetic parameters, PDCZ.m. exhibits a hyperbolic v/S-plot [117], and is not influenced by phosphate [76, 82], pyruvate, and pyruvamide as effectors. A recent investigation compared the decarboxylation of pyruvate by PDCS.c. and PDCZ.m. by measuring isotopic effects [117]. The results clearly show that PDCZ.m. exhibits the highest flux, and the greatest amount of stabilization for all transition states along the reaction path. This implies that the introduction of the regulatory apparatus in PDCS.c. has been accompanied by a decrease in catalytic power.

5 Application of PDC in Biotransformations

5.1 Substrate Spectrum of the Decarboxylation Reaction

Various biotransformations concerning the decarboxylation of α-keto acids and the formation of α-hydroxy ketones with different aldehydes as cosubstrates have been performed with either whole cell-systems, mainly fermenting yeast [116, 143–151], or isolated enzymes from wheat germ [27, 33, 120, 152], yeast

[115, 118, 151, 153, 155–157] and *Z. mobilis* [118, 153–157]. These studies revealed relatively broad substrate spectra with respect to the decarboxylation reaction. Besides pyruvate, PDC*S.c.* accepts longer aliphatic α-keto acids, like α-ketobutyric acid, α-ketovaleric acid, and branched aliphatic α-keto acids, like α-ketoisocarproic acid and α-ketoisovalereric acid [33, 115, 116, 118]. Comparative studies of the enzymes from yeast and *Z. mobilis* established a lower affinity of the bacterial enzyme for longer-chain aliphatic α-keto acids [118].

Various 3-halopyruvate derivatives have been reported to be decarboxylated and simultaneously dehalogenated, yielding acetate, carbon dioxide and a halogenide ion as the only reaction products [133, 158]. A similar reaction may occur upon enzymatic decarboxylation of 3-hydroxypyruvate, which is an alternative substrate and a strong competitive inhibitor of PDC [159].

The decarboxylation of glyoxalate results in the formation of "active formaldehyde". Since the release of formaldehyde is very slow, glyoxalate is a uncompetitive inhibitor of PDC [128, 160].

Even benzoylformate and some derivatives may be decarboxylated by PDC*S.c.* [115].

5.2 Formation of α-Hydroxy Ketones

As demonstrated in Scheme 3, the ThDP-bound "active aldehyde" **6** as an acyl-donor may be added to a second aldehyde cosubstrate (acyl-acceptor) in an acyloin condensation-type reaction. This carboligase reaction was intensively investigated with acetaldehyde as an acyl-donor, which may be either condensed to a further acetaldehyde molecule yielding acetoin [1, 26, 27, 29, 63, 153, 154] or to a wide range of various aliphatic, aromatic and heterocyclic aldehydes [5, 14, 118, 151, 154–157, 161].

Apart from the large number of different α-keto acids which may be decarboxylated by PDC, only a few of the resulting aldehydes may be transferred to a second aldehyde molecule to form an α-hydroxy ketone [151]. Besides acetaldehyde, which is the best acylanion equivalent, propionaldehyde and butyraldehyde have been condensed to benzaldehyde by baker's yeast after decarboxylation of the corresponding α-keto acids [116, 149].

5.3 Stereo-Control of the Carboligase Reaction

A detailed investigation of the carboligase reaction mediated by PDC from yeast, wheat germ and *Z. mobilis* revealed that the stereo-control of this reaction is only strict with aromatic aldehydes as acylanion-acceptors, while the formation of acetoin (3-hydroxybutan-2-one) resulted in mixtures of the (*R*)- and (*S*)-enantiomer.

In contrast to PDC from wheat germ and PDCS.c., which produce an enantiomeric excess of about 50% of (R)-(+)-acetoin [26, 27, 63, 162], PDCZ.m. synthesizes predominantly the (S)-(−)-acetoin [153]. Whereas the acylation of aromatic aldehydes by either PDC from yeast or Z. mobilis gives exclusively the (R)-acyloins [149, 154, 163, 164].

These differences in the control of the product stereochemistry have recently been investigated by molecular modeling techniques [60, 154]. From these studies, the relevance of the side-chain of isoleucine 476 (PDCS.c.) (Table 2) for the stereo-control during the formation of aromatic α-hydroxy ketones became obvious, since this side-chain may protect one site of the α-carbanion/enamine **6** (Scheme 3) against the bulky aromatic cosubstrate. Nevertheless, the smaller methyl group of acetaldehyde can bind to both sites of the α-carbanion/en-amine. The preference for one of the two acetoin enantiomers has been interpre-tated in terms of different Boltzmann distributions between the two binding modes of the bound acetaldehyde [155].

6 Site-Directed Mutagenesis of Pyruvate Decarboxylase

Site-directed mutagenesis has become a common tool for elucidating reaction mechanisms. The method has also been applied for the change of substrate- and cofactor specificities [165–169]. In contrast to the principles of the "directed evolution of enzymes", which is the subject of the review by Frances Arnold in this issue, the so-called "rational approach" to modify proteins re-quires considerable knowledge of the structure and the reaction mechanisms. Nevertheless, the effects of single side-chain mutations have to be interpreted cautiously, as long as no structural information on the mutant enzymes is available, since the influence of the mutation may not only be limited to the direct environment of the side-chain. A detailed consideration of the strategies and techniques involved in protein engineering is beyond the scope of this article. A recent monograph summarizes the relevant methods in molecular biology, computer modeling and structure investigation [166, 167, 169].

Protein design by site-directed mutagenesis on pyruvate decarboxylase became possible after the 3D-structure of the enzyme from *Saccharomyces uvarum* had become available [35]. Based on sequence comparison and second-ary structure prediction, the 3D-structure of the yeast enzyme served as a model for PDCZ.m. [163]. The point mutations which have been introduced into the two enzymes (Tables 4 and 5) concern catalytically important residues as well as significant side-chain interactions at the domain interface of the dimer. Besides, site-directed mutagenesis offered a powerful tool to improve the car-boligase reaction of PDCZ.m. with respect to the synthesis of (R)-PAC [163, 164, 170].

Table 4. Various mutations of PDC1 from *Saccharomyces cerevisiae*

Mutation	V_{max} (U/mg)	$S_{0.5}$ (mM)	Hill-coeff. n_h	v/S-plot	Ref.
wt-PDCS.c.	45–55	1.0	2.0–2.2	sigmoid.	[34, 40–47] [62–63, 70] [75, 77–80] [83, 107–116] [182]
Asp 28 Ala	7.8	n.d.	1.7	sigmoid.	[172]
Glu 51 Gln	0.3	n.d.	2.0		[172]
Asp	0.5	n.d.	2.2		
Asn	0.6	n.d.	2.1		
Ala	0.11	n.d.	1.1		
His 92 Gly	25	n.d.	1.2		[172, 182]
Ala	25	10.1 ± 0.5	1.32 ± 0.5		
Lys	25	n.d.	1.7		
Cys 152 Ala	40–45	3.23 ± 0.5	1.98 ± 0.01		[172, 182]
Cys 221 Ser	4.8–15	$4.0/7.6 \pm 0.5$	0.8–1.0	hyperbol.	[138, 172, 182]
Cys 222 Ser	30–48	$1.0/4.3 \pm 0.6$	1.6–1.9	sigmoid.	[138, 172, 182]
Cys 221 Ser/ Cys 222 Ser	5–13	8.8 ± 0.4	0.98 ± 0.02	hyperbol.	[138, 182]
Ile 415 Leu	3.1	n.d.	1.7		[172]
Val	8.3	n.d.	2.2		
Thr	1.6	n.d.	2.2		
Glu 477 Asp	0.85	n.d.	1.3		[172]

6.1 Mutations of Catalytically Important Amino Acids

The relevance of Glu50 in PDCZ.*m*. [171], which corresponds to Glu51 in PDCS.*c*. [172] has been demonstrated by site-directed mutagenesis to Gln, Asp and Ala. All mutations resulted in a greatly diminished specific activity, and a decreased affinity for the cofactors ThDP and Mg^{2+} (Tables 4, 5)

The X-ray structure of PDC from yeast revealed a hydrogen bond between the side-chain of Glu50 and N1'-ThDP **3** (Scheme 2). This interaction favours the amino-imino-tautomerization reaction and thus enables the 4'NH$_2$-group to react as a proton acceptor or donor [35, 51]. The "V"-conformation of enzyme-bound ThDP brings the 4'NH$_2$-group into interactive contact with C2-ThDP. This is stabilized by the side-chain of Ile415 (Table 1). A diminution in the size of the group at 415 leads to disminished specific activity (Table 4) [172].

Further mutations concerned the ThDP-binding motif GDG-(X)$_{24}$-NN (Figs. 1 and 2, Scheme 5) [172, 173], which is located in the γ-domain of the PDC chain and contains two amino acid residues, which are involved into the coordination of the Mg^{2+}-cofactor binding (Asp 444/440, Asn 471/467), and Ser 446/442, which is involved in the coordination of the diphosphate part of ThDP [35] (Scheme 5).

```
            *   *  + +     +     +       +                    *
PDC S.c.  443G D G S L Q L T V Q E I S T M I R W G L K P Y L F V L N N471
PDC Z.m.  339G D G S F Q L T A Q E V A Q M V R L K L P V I I F L I N N467
```

Scheme 5. Comparison of the ThDP-binding motif of PDCS.c. and PDCZ.m. [173]. The amino acid residues involved in the binding of cofactors are marked with *. Those which are involved in polar subunit-subunit-interactions within the dimer are marked with + [51]. Amino acids which have been investigated by site-directed mutagenesis are underlined (Tables 4 and 5)

Table 5. Various mutations of PDC from *Zymomonas mobilis*

Mutation	V_{max} (U/mg)	$S_{0.5}$ (mM)	further effects	Ref.
Z. mobilis wt-PDCZ.m.	120–180	0.4–1.1	n_h 1.0 – v/S-plot hyperbolic	[72, 76, 82, 106, 117-119]
Glu 50 Gln	0.32	n.d.		[171]
Asp	2.0	0.5		[175]
Thr 119 Ile	29	n.d.	– only stable in presence of the cofactors	[174]
Thr 132 Ala	6.4	0.53	– very instable	[174]
Asp 440 Asn	inactive		– no ThDP binding	[176]
Thr	inactive			
Gly	inactive			
Glu	14.2	0.95	· dimer under non-turnover conditions	
Glu 449 Asp	12.5	0.52	– similar to wt-PDC	[174]
Pro 459 Gly	13.0	0.87	– all Pro 459 mutants are	[174]
Ala	28.2	0.67	similar to wt-PDC	
Asn 467 Asp	8.4	0.95	– lower affinity for ThDP	[174]
Gln	inactive		– no binding of ThDP/Mg^{2+}	
Trp 487 Leu	n.d.	0.8	– reduced cofactor binding n_h 0.99	[176]
His 113 Gln	inactive		– reduced cofactor binding	[175]
Phe 496 Leu	52.3	0.97	– all Phe 496 mutants have higher affinity for ThDP	[171]
Ile	22.6	1.11	and a small decrease in affinity for Mg^{2+}	[175]
His	28.8	1.06		
Trp 392 Ala	65	3.8	· improved synthesis of	[163]
Ile	70	4.3	(R)-PAC	
Met	80	6.7		[164]
Phe	80	5.4		
Asn	60	4.0		[170]

Consequently, for mutant enzymes with exchanges in position Asp440 and Asn467, the ability to bind the cofactors was greatly diminished. All amino acid exchanges in this area resulted in a high degree of desactivation [171, 174].

Another mechanistically important position concerned the substrate activation site of PDCS.c. Cys221 has unequivocally been identified as the site of substrate activation by site-directed mutagenesis [138]. The fact that Cys221 is found > 20 Å away from the ThDP-site raised the question of how the information is relayed from Cys221 to the coenzyme. A chemically plausible scenario is that the Cys nucleophile forms a tetrahedral hemithioketal adduct with the α-keto group of the substrate, or with pyruvamide [51, 135]. A computer model construction of such an adduct revealed unfavorably short contacts with His92. Thus either Cys221, His92 or both must move to accommodate the adduct. This structural distortion may be propagated by adjacent residues to the ThDP site. This model of substrate activation is supported by the mutation of His92 to Gly, Ala or Lys, which greatly diminished the Hill coefficient compared to wt-PDCS.c. (Table 4) [51, 172, 182].

Nevertheless, this model may only explain the substrate activation behaviour of PDC from various yeasts (Saccharomyces cerevisae, Klyveromyces marxianus, Klyveromyces lactis, Hanseniaspora uvarum) since these are the only PDC-sequence with a cysteine in position 221 and a histidine in position 95. All other available PDC sequence contain a lysine residue in the position corresponding to Cys221 and an asparagine in the position corresponding to His92 (Figs. 1 and 2), whether they show substrate activation behaviour (like PDC from pea, wheat germ and rice) or not (PDCZ.m.). The only exception is PDC from Aspergillus parasiticus with a lysine and glutamine acid in the positions corresponding to His95 and Cys221, respectively (Fig. 2). Thus, different routes of substrate activation, independent of a hemithioketalformation thioketal formation, must exist.

6.2 Mutations to Elucidate Special Side-Chain Interactions

The intrinsic tryptophane fluorescence of PDC is quenched upon binding of the cofactors Mg^{2+} and ThDP [72, 175]. This observation led to the conclusion that one or more Trp-residues are involved in the binding of ThDP by the enzyme. Before the 3D-structure of PDC was published, the substitution of a Trp-residue, which is conserved among PDC-sequences from various species (Trp487 in PDCZ.m. corresponding to Trp493 in PDCS.c.) by Leu was reported, which resulted in a reduction of the quenching effect [177]. Inspection of the 3D-structure revealed that this Trp-residue is located at the interface between the two monomers forming one dimer and not in the direct environment of ThDP [35, 51]. It was proposed that the observed quenching effect upon binding of the cofactors to the active center is a consequence of altering the ring stacking with a phenylalanine from the adjacent subunit (Phe496 in PDCZ.m., Phe502 in PDCS.c.) [35]. This assumption could not be confirmed by site-directed mutagenesis studies [171]. The mutation of Phe496 to Leu, Ile or His did not eliminate the fluorescence quenching effect upon addition of the cofactors. Thus, a specific interaction of Trp487 with Phe496 could be excluded.

The γ-carboxyl group of Glu477 in PDCS.c. (Glu473 in PDCZ.m.) is assumed to be involved in the stabilization of ThDP-bound reaction intermediates [60] (Table 2). A mutant enzyme with Asp in an equivalent position exhibits only minor decarboxylase activity [172].

6.3 Mutations to Optimize the Carboligase Reaction of PDCZ.m.

The application of isolated enzymes in biotransformations is often superior to the use of cell-systems, since the various enzymes existing in living cells may result in the formation of by-products and thus cause side-reactions of the substrates and impair the yield of the desired products.

During the biotransformation of benzaldehyde by fermenting yeast (Scheme 1) to (R)-PAC the reduction of the benzaldehyde to benzyl alcohol is a serious problem, caused partially by the presence of alcohol dehydrogenases

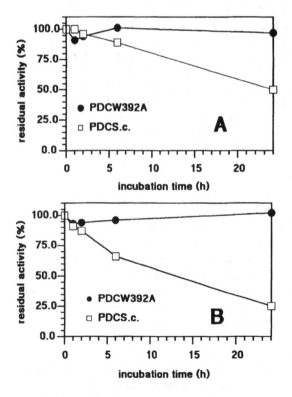

Fig. 3. Stability of wt-PDCS.c. and the mutant enzyme PDCZ.m.-W392A in 50 mM Mes/KOH-buffer, 5 mM MgSO$_4$, 0.1 mM ThDP at **A** 25 °C and **B** 30 °C [170]

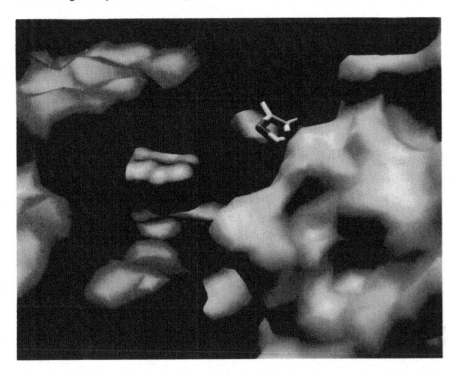

Fig. 4. Computer graphics of the domain interface building the channel to the active center in PDC. The thiazolium ring of ThDP is visible at the bottom of the channel. A tryptophane residue (blue) has been engineered into the crystal structure of PDCS.*u.* by means of computer graphics. The picture was generated by J. Grötzinger using the program GRASP [181]

in the yeast [10, 15, 161]. Further by-products are acetoin, butane-2,3-dione (diketone), 1-phenylpropan-2,3-dione, 1-phenyl-1,2-propandiol, benzoic acid and 2-hydroxypropiophenone [177–179]. These problems may be overcome by the application of isolated enzymes in appropriate enzyme reactors [180].

The high stability and high catalytic activity makes PDCZ.*m.* well suited for the application in biotransformations. Unfortunately, this enzyme is a less efficient catalyst with respect to the formation of PAC compared to PDCS.*c.* [118]. On the other hand, PDCS.*c.* is not suited for biotransformation in enzyme reactors due to its low stability (Fig. 3). Attempts to improve the synthetic capacity of PDCZ.*m.* for the synthesis of (R)-PAC started with an investigation of the deep cleft leading to the active center by means of computer graphics, based on the 3D-structure of PDCS.*u.* and sequence alignments [163, 164]. These studies revealed a Trp-residue in position 392 of PDCZ.*m.*, which is replaced by Ala in PDCS.*u.* (Fig. 1, 2, 4). The assumption that this bulky Trp-residue may hinder the passage of aromatic molecules such as benzaldehyde and PAC was proved by site-directed mutagenesis studies [163, 164, 170]. The mutant enzyme PDC W392A has, by a factor of 3–4, higher carboligase activity

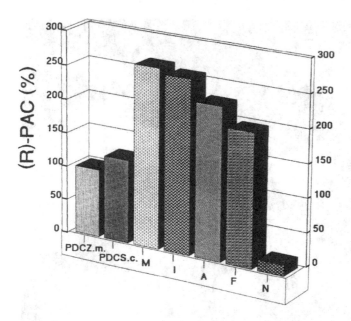

Fig. 5. Enzymatic synthesis of (R)-PAC catalysed by PDCZ.*m*., PDC*S.c.* and various mutants of PDCZ.*m*. in position 392 in a simplified batch reaction [170] (The amino acids which replace the original Trp-residue are indicated, respectively).

with respect to the formation of (R)-PAC using a batch-reactor with a coupled enzymatic system to remove acteladehyde [163–164]. The sequence space in this position has further been investigated by multiple replacements using site-directed mutagenesis [170] (Table 5). The substitution of Trp is connected with a decrease of the decarboxylase activity to about 50% of the wt-activity. The km-values for pyruvate of the mutant enzymes are by a factor of 3–7 higher compared to the wt-enzyme. The stability parallels the activity profile of the mutant enzymes and decreased in the series Trp > Phe = Met > Ile > Ala ≥ Asn. Although the stability of the mutant enzymes is slightly diminished compared to wt-PDCZ.*m*., they are all more active and more stable than wt-PDC*S.c.* (Table 3a, Fig. 3).

There was no difference observed between wt-PDCZ.*m*. and the mutant enzymes with respect to the formation of acetoin. However, the biotransformation of benzaldehyde to (R)-PAC **1** was best performed with the mutant enzymes PDCW392M and PDCW392I (Fig. 5). As a consequence of these studies, Trp392 in PDCZ.*m*. could be characterized as a further key residue which is responsible for the differences between PDC*S.c.* and PDCZ.*m*.. Besides the absence of a regulatory apparatus in PDCZ.*m*., which was shown to impair the catalytic power of PDC*S.c.* [117], the Trp-residue is partially responsible for the higher stability and also for the higher activity of the bacterial enzyme.

The improvement of the carboligase reaction of PDCZ.*m*. is a further example for the potency of the "rational" approach in designing enzymes with special properties. In this case the design was successful although the 3D-structure of PDCZ.*m.* is not yet available and the mutants had to be planned based on the 3D-structure of PDCS.*c.*. The success of this approach is an indirect proof that the two enzymes are structurally similar, especially at the cofactor binding site and the monomer-monomer interface (Tables 1 and 2).

7 References

1. Neuberg C, Karczag L (1911) Biochem Zeitschr 36: 68
2. Neuberg C, Karczag L (1911) Biochem Zeitschr 36: 76
3. Neuberg C, Rosenthal P (1913) Biochem Zeitschr 51: 128
4. Neuberg C, Hirsch J (1921) Biochem Zeitschr 115: 282
5. Neuberg C, Ohle H (1922) Biochem Zeitschr 127: 327
6. Hildebrandt G, Klavehn W (1932) German Patent No. 548 459.
7. Hildebrandt G, Klavehn W (1934) U.S. Patent 1,956,950.
8. Pfanz H, Heise D (1957) German Patent No. 13683.
9. Selly RJ, Heefner DL, Hageman RV, Yarus MJ, Sullivan SA (1994) U.S. Patent, No. 5,312,742.
10. Nikolova P, Ward OP (1991) Biotechnol Bioeng 20: 493
11. Nikolova P, Ward OP (1994) Biotechnol Lett 16: 7
12. Nikolova P, Long A, Ward OP (1991) Biotechnol Techniques 5: 31
13. Long A, Ward OP (1989) J Ind Microbiol 4: 49
14. Long A, Ward OP (1989) Biotechnol Bioeng 34: 933
15. Mahmoud WM, El-Sayed ABMM, Coughlin RW (1990) Biotechnol Bioeng 36: 256
16. Mahmoud WM, El-Sayed AHMM, Coughlin RW (1990) Biotechnol Bioeng 36: 47
17. Mahmoud WM, El-Sayed AHMM, Coughlin RW (1990) Biotechnol Bioeng 36: 55
18. Agarwal SC, Basu SK, Vora VC, Mason JR, Pirt SJ (1987) Biotechnol Bioeng 29: 783
19. Chow YS, Shin HS, Adesina AA, Rogers PL (1995) Biotechnol Lett 17: 1201
20. Liew MKH, Fane AG, Rogers PL (1995) Biotechnol Bioeng 48: 108
21. Shin HS, Rogers PL (1995) Appl Microbiol Biotechnol 44: 7
22. Shin HS, Rogers PL (1996) Biotechnol Bioeng 49: 429
23. Shin HS, Rogers PL (1996) Biotechnol Bioeng 49: 52
24. Dirscherl W (1931) Hoppe-Seyler's Z Physiol Chem 201: 47
25. Langenbeck W, Wrede H, Schlockermann W (1934) Hoppe-Seyler's Z Physiol Chem 227: 263
26. Singer TP, Pensky J (1951) Arch Biochem Biophys 31: 457
27. Singer TP, Pensky J (1952) Arch Biochem Biophys Acta 9: 316
28. Juni E (1952) J Biol Chem 195: 727
29. Juni E (1961) J Biol Chem 236: 2302
30. Lohmann K, Schuster P (1931) Naturwissenschaften 19: 18
31. Auhagen E, Hartmann M, von Laue M, Rosenheim A, Volmer M (1931) Naturwissenschaften 19: 916
32. Lohmann K, Schuster P (1937) Biochem Zeitschr 294: 188
33. Green DE, Herbert D, Subrahmanyan V (1941) J Biol Chem 138: 327
34. Flatau S, Fischer G, Kleinpeter E, Schellenberger A (1988) FEBS Lett 233: 379
35. Dyda F, Furey W, Swaminathan S, Sax M, Farrenkopf BC, Jordan F (1993) Biochemistry 32: 6165
36. Breslow R (1958) J Am Chem Soc 80: 3719
37. Breslow R, McNelis E (1959) J Am Chem Soc 81: 3080
38. Krampitz LO, Greull G, Miller CS, Bicking JB, Skeggs HR, Sprangue JM (1958) J Am Chem Soc 80: 5893
39. Holzer H, da Fonzea-Wollheim F, Kohlhaw G, Woenckhaus CW (1961) Ann NY Acad Sci 98: 453
40. Schellenberger A (1967) Angew Chem 79: 1050

41. Schellenberger A, Müller V, Winter K, Hübner G (1966) Hoppe-Seyler's Z Physiol Chem 344: 244
42. Wittorf JH, Gubler CJ (1970) Eur J Biochem 14: 53
43. Ullrich J, Donner I (1970) Hoppe-Seyler's Z Physiol Chem 351: 1030
44. Ullrich J, Donner I (1970) Hoppe-Seyler's Z Physiol Chem 351: 1026
45. Kuo DJ, Jordan F (1983) Biochemistry 22: 3735
46. Kuo DJ, Jordan F (1983) J Biol Chem 225: 13415
47. Hübner G, Atanassova M, Schellenberger A (1986) Biomed Biochem Acta 45: 823
48. Kluger R (1987) Chem Rev 87: 863
49. Schellenberger A (1990) Chem Ber 123: 1489
50. Schellenberger A, Neef H, Golbig R, Hübner G, König S (1990) Mechanistic aspects of thiamine pyrophosphate enzymes via site-directed substitutions of the coenzyme structure. In: Bisswanger H, Ullrich H (eds) Biochemistry and physiology of thiamin diphosphate enzymes. VCH, Weinheim, p 3
51. Arjunan P, Umland T, Dyda F, Swaminathan S, Furey W, Sax M, Farrenkopf B, Gao Y, Zhang D, Jordan F (1996) J Mol Biol 256: 590
52. Alvarez FJ, Ermer J, Hübner G, Schellenberger A, Schowen RL (1991) J Am Chem Soc 113: 8402
53. Schellenberger A, Hübner G (1967) Hoppe-Seyler's Z Physiol Chem 348: 491
54. Crane EJ, Vaccaro JA, Washabaugh MW (1993) J Am Chem Soc 115: 8912
55. Stivers JT, Washabaugh MW (1993) Biochemistry 32: 13472
56. Harris TK, Washabaugh MW (1995) Biochemistry 34: 13994
57. Harris TK, Washabaugh MW (1995) Biochemistry 34: 14001
58. Lindqvist Y, Schneider G (1993) Curr Opin Struct Biol 3: 896
59. Muller YA, Lindquist Y, Furey W, Schulz GE, Jordan F, Schneider G (1993) Structure 1: 95
60. Lobell M, Crout DHG (1996) J Am Chem Soc 118: 1867
61. Kern D, Kern G, Neef H, Killenberg-Jabs M, Tittmann K, Hübner G (1996) How thiamin is activated in enzymes (Bisswanger H, Schellenberger A, eds) Biochemistry and physiology of thiamin diphosphate enzymes, A. & C. Intenmann, Wissenschaftlicher Verlag, Prien, p 25
62. Ermer J, Schellenberger A, Hübner G (1992) FEBS Lett 299: 163
63. Chen GC, Jordan F (1984) Biochemistry 23: 3582
64. Zeng X, Chung A, Haran M, Jordan F (1991) J Am Chem Soc 113: 5842
65. Menon-Rudolph S, Nishikawa S, Zeng X, Jordan F (1992) J Am Chem Soc 114: 10110
66. Lindqvist Y, Schneider , Ermler U, Sundström M (1992) EMBO J 11: 1: 2373
67. Muller YA, Schulz GE (1993) Science 259: 965
68. Muller YA, Schumacher G, Rudolph R, Schulz GE (1994) J Mol Biol 237: 315
69. Kubowitz F, Lüttgens W (1941) Biochem Zeitschr 307: 170
70. Morey AV, Juni E (1968) J Biol Chem 243: 3009
71. Langston-Unkefer PJ, Lee TC (1985) Plant Physiol 79: 436
72. Diefenbach RJ, Duggleby RG (1991) Biochem J 276: 439
73. Lautens JC, Kluger R (1992) J Org Chem 57: 6410
74. Vaccaro JA, Crane EJ, Harris TH, Washabaugh MW (1995) Biochemistry 34: 12636
75. Juni E, Heym GA (1968) Arch Biochem Biophys 127: 79
76. Pohl M, Mesch K, Rodenbrock A, Kula MR (1995) Biotechnol Appl Biochem 22: 95
77. Eppendorfer S, König S, Golbik R, Neef H, Lehle K, Jaenicke R, Schellenberger A, Hübner G (1993) Hoppe-Seyler's Z Physiol Chem 374: 1129
78. Gounaris AD, Turkenkopf I, Buckwald S, Young A (1971) J Biol Chem 246: 1302
79. Gounaris AD, Turkenkopf I, Civerchia LL, Greenlie J (1975) Biochim Biophys Acta 405: 492
80. König S, Svergun D, Koch MHJ, Hübner G, Schellenberger A (1992) Biochemistry 31: 8726
81. König S, Svergun D, Koch MHJ (1996) Comparison of subunit interactions of pyruvate decarboxylase from different organisms by small-angle solution scattering. In: Bisswanger H, Schellenberger A (eds) Biochemistry and physiology of thiamin diphosphate enzymes. A. & C. Intemann, Wissenschaftlicher Verlag, Prien, p 160
82. Pohl M, Grötzinger J, Wollmer A, Kula MR (1994) Eur J Biochem 224: 651
83. Hopmann RFW (1980) Eur J Biochem 110: 311
84. Killenberg-Jabs M, Hübner G, Kern G (1996) Folding and unfolding of homotetrameric pyruvate decarboxylase from S. cerevisiae. In: Bisswanger H, Schellenberger A (eds) Biochemstry and physiology of thiamin diphosphate enzymes. A. & C. Intemann, Wissenschaftlicher Verlag, Prien, p 195

85. Perata P, Alpi A (1993) Plant Science 93: 1
86. Gancedo C, Serrano R (1989) Energy-yielding metabolism. In: Rose AH, Harrisin JS (eds) The yeast, vol. 3. Academic Press, New York, p 205
87. Kellermann E, Seeboth PG, Hollenberg CP (1986) Nucleic Acids Res 14: 8963
88. Hohmann S (1991) J Bacteriol 173: 7963
89. Hohmann S (1991) Curr Genet 20: 373
90. Hohmann S, Cederberg H (1990) Eur J Biochem 188: 615
91. Holloway P, Subden RE (1994) Yeast 10: 1581
92. Holloway P, Subden RE (1993) Curr Genet 24: 274
93. Bianachi MM, Tizzani L, Destruelle M, Frontali L, Wesolowski-Louvel M (1996) Molec Microbiol 19: 27
94. Alvarez ME, Rosa AL, Temperoni ED, Wolstenholme A, Panzetta G, Patrita L, Maccioni HJF (1993) Gene 130: 253
95. Sanchis V, Vinas I, Roberts IN, Jeenes DJ, Watson AJ, Archer DB (1994) FEMS Microbiol Lett 117: 207
96. Kelley PM (1989) Plant Molec Biol 13: 213
97. Kelley PM, Godfrey K, Lal SK, Alleman M (1991) Plant Molec Biol 17: 1259
98. Hossain MA, Hug E, Hodges TK (1994) Plant Physiol 106: 799
99. Foster A, Chase T (1995) FASEB J 9: A1292
100. Foster A, Chase T (1995) Plant Physiol 108(2.Supp) : 75
101. Mücke U, Wohlfarth T, Fiedler U, Bäumlein H, Rücknagel KP, König S (1996) Eur J Biochem 237: 373
102. Bucher M, Brander K, Sbicego S, Mandel T, Kühlemeyer C (1995) Plant Molec Biol 28: 739
103. Neale AD, Scopes RK, Wettenhall REH, Hoogenraad NJ (1987) Nucleic Acids Res 15: 1753
104. Conway T, Osman YA, Konnan JI, Hoffmann EM, Ingram LO (1987) J Bacteriol 169: 949
105. Reynen M, Sahm H (1988) J Bacteriol 170: 3310
106. Miczka G, Vernau J, Kula MR, Hofman B, Schomberg D (1992) Biotechnol Appl Biochem 15: 192
107. Alvarez FJ, Ermer J, Hübner G, Schellenberger A, Schowen RL (1995) J Am Chem Soc 117: 1678
108. Farrenkopf BC, Jordan F (1992) Protein Express Purif 3: 101
109. Hübner G, König S, Schellenberger A, Koch MHJ (1990) FEBS Lett 266: 17
110. Hübner G, König S, Schnackerz KD (1992) FEBS Lett 314: 101
111. Jordan F, Dikdan G (1995) FASEB J 9: A1285
112. König S, Hübner G, Schellenberger A (1990) Biomed Biochem Acta 6: 465
113. König S, Svergun D, Koch MHJ, Hübner G, Schellenberger A (1993) Eur Biophys J 22: 185
114. Kuo DJ, Dikdan G, Jordan F (1986) J Biol Chem 261: 3316
115. Lehmann H, Fischer G, Hübner G, Kohnert KD, Schellenberger A (1973) Eur J Biochem 32: 83
116. Suomalainen H, Linnehalme T (1966) Arch Biochem Biophys 114: 502
117. Sun S, Duggleby RG, Schowen RI (1995) J Am Chem Soc 117: 7317
118. Bringer-Meyer S, Sahm H (1988) Biocatalysis 1: 321
119. Neale AD, Scopes RK, Wettenhall REH, Hoogenraad NJ (1987) J Bacteriol 169: 1024
120. Crout DHG, Littlechild J, Morrey SM (1986) J Chem Soc Perkin Trans I: 105
121. Kluger R, Gish G, Kauffman G (1984) J Biol Chem 259: 8960
122. Ullrich J (1982) Ann N Y Acad Sci 378: 287
123. Zehender H, Ullrich J (1985) FEBS Lett 180: 51
124. Zehender H, Trescher D, Ullrich J (1987) Eur J Biochem 167: 149
125. Mücke U, König S, Hübner G (1995) Biol Chem Hoppe-Seyler 376: 111
126. Lebova S (1990) Characterization of pyruvate decarboxylases isolated from germinating seeds. In: Bisswanger H, Ullrich J (eds) Biochemistry and physiology of thiamin diphosphate enzymes. VCH, Weinheim, p 133
127. Lee CL, Langston-Unkefer PJ (1979) Plant Physiol 79: 242
128. Rivoal J, Ricard B, Pradet A (1990) Eur J Biochem 194: 791
129. Oba K, Uritani I (1975) J Biochem (Tokyo) 77: 1205
130. Oba K, Uritani I (1982) Pyruvate decarboxylase from sweet potato roots. In: Wood WA (ed) Methods in Enzymology, vol. 90. Academic Press, New York, p 528
131. Lowe SE, Zeikus JG (1992) J Gen Microbiol 138: 803
132. Bioteux A, Hess B (1970) FEBS Lett 9: 293
133. Gish G, Smyth T, Kluger R (1988) J Am Chem Soc 110: 6230

134. Schellenberger A, Hübner G (1970) Z Chem 10: 436
135. Hübner G, Weidhase R, Schellenberger A (1978) Eur J Biochem 92: 175
136. Hübner G, König S, Schellenberger A (1988) Biomed Biochem Acta 1: 9
137. Zeng X, Farrenkopf BC, Hohmann S, Dyda F, Furey W, Jordan F (1993) Biochemistry 32: 2704
138. Baburina I, Gao Y, Hu Z, Jordan F, Hohmann S, Furey W (1994) Biochemistry 33: 5630
139. Schaaf I, Green JBA, Gozalbo D, Hohmann S (1989) Curr Genet 15: 75
140. Seeboth PG, Bohnsack K, Hollenberg CP (1990) J Bacteriol 172: 678
141. Candy JM, Duggleby RG, Mattick JS (1991) J Gen Microbiol 137: 2811
142. Ullrich J, Leuble (1986) Hoppe-Seyler's Z Physiol Chem 367 Suppl: 363
143. Abraham WR, Stumpf B (1987) Z Naturforsch C Biosci 42c: 559
144. Cardillo R, Servi St, Tinti C (1991) Appl Microbiol Biotechnol 36: 300
145. Ebert C, Gardossi T, Gianferrara T, Linda P, Morandini C (1994) Biocatalysis 10: 15
146. Fuganti C, Grasselli P, Poli G, Servi S, Zorella A (1988) J Chem Soc Chem Comm: 1619
147. Fuganti C, Grasselli P, Servi St, Spreafico F, Zirotti C (1984) J Org Chem 49: 4087
148. Fuganti C, Grasselli P, Spreafico F, Zirotti C (1984) J Org Chem 49: 543
149. Fuganti C, Grasselli P, Poli G, Servi S, Höberg HE (1988) J Chem Soc Perkin Trans I: 3061
150. Ohta H, Ozaki K, Konishi J, Tsuchihashi G (1986) Agric Biol Chem 50: 1261
151. Csuk R, Glänzer BI (1991) Chem Rev 91: 49
152. Singer TP, Pensky J (1952) J Biol Chem 196: 375
153. Bornemann S, Crout DHG, Dalton H, Hutchinson DW, Dean G, Thomson N, Turner MM (1993) J Chem Soc Perkin Trans I: 309
154. Bornemann S, Crout DHG, Dalton H, Kren V, Lobell M, Dean G, Thomson N, Turner MM (1996) J Chem Soc Perkin Trans I: 425
155. Crout DHG, Dalton H, Hutchinson DW, Miyagoshi M (1991) J Chem Soc Perkin Trans I: 1329
156. Crout DHG, Davies S, Heath RJ, Miles CO, Rathbone DR, Swoboda BEP (1994) Biocatalysis 9: 1
157. Kren V, Crout DHG, Dalton H, Hutchinson DW, König W, Turner MM, Dean G, Thomson N (1993) J Chem Soc Chem Commun: 341
158. Schellenberger A, Hübner G, Atanassowa M, Sieber M, König S (1986) Wiss Z Univ Halle 3: 37
159. Thomas G, Diefenbach RJ, Duggleby RG (1990) Biochem J 266: 305
160. Uhlemann H, Schellenberger A (1976) FEBS Lett 63: 37
161. Long A, James P, Ward OP (1989) Biotechnol Bioeng 33: 657
162. Singer TP (1952) Biochim Biophys Acta 8: 108
163. Bruhn H, Pohl M, Grötzinger J, Kula MR (1995) Eur J Biochem 234: 650
164. Bruhn H, Pohl M, Mesch K, Kula MR (1996) Deutsche Patentanmeldung 195 23 269. 0–41
165. Fagain CO (1995) Biochim Biophys Acta 1252: 1
166. Hahn U, Heinemann U (1994) Structure determination, modeling and site-directed mutagenesis. In: Wrede P, Schneider G (eds) Concepts in protein engineering and design. De Gruyter, Berlin, p 109
167. Schomburg D (1994) Rational protein design of proteins with new properties. In: Wrede P, Schneider G (eds) Concepts in protein engineering and design. De Gruyter, Berlin, p 169
168. Fersht AR, Winter G (1986) TIBS 17: 292
169. Gaddis TJ, Oxender DL (1994) An Introduction of protein engineering. In: Wrede P, Schneider G (eds) Concepts in protein engineering and design. De Gruyter, Berlin, p 1
170. Pohl M, Mesch K, Bruhn H, Goetz G, Iding H, Kula MR (1996) Site-directed mutagenesis studies on the pyruvate decarboxylase from Zymomonas mobilis In: (Bisswanger H, Schellenberger A, eds) Biochemistry and physiology of thiamin diphosphate enzymes, A. & C. Intemann, Wissenschaftlicher Verlag, Prien, p 160
171. Candy JM, Koga J, Nixon PF, Duggleby RG (1996) Biochem J 315: 745
172. Jordan F, Barburina I, Gao Y, Kahyaoglu A, Nemeria N, Volkov A, Yi JZ, Zhang D, Machado R, Guest JR, Furey W, Hohmann S (1996) New insights to the regulation of thiamin diphosphate decarboxylases by substrate and ThDP.Mg(II) In: (Bisswanger H, Schellenberger A, eds) Biochemistry and physiology of thiamin diphosphate enzymes. A. & C. Intemann, Wissenschaftlicher Verlag, Prien, p 53
173. Hawkins CF, Borges A, Perham RN (1989) FEBS Lett 255: 77
174. Candy JM, Duggleby RG (1994) Biochem J 330: 7

175. Candy JM, Nixon PF, England R, Schenk G, Koga J, Duggleby RG (1996) Site-directed mutagenesis of E50, F496 and His113 in *Zymomonas mobilis* pyruvate decarboxylase In: (Bisswanger H, Schellenberger A, eds) Biochemistry and physiology of thiamin diphosphate enzymes, A. & C. Intemann, Wissenschaftlicher Verlag, Prien, p 82
176. Diefenbach RJ, Candy JM, Mattick JS, Duggleby RG (1992) FEBS Lett 296: 95
177. Becvarova H, Hanc O (1963) Folia Microbiol 8: 42
178. Voets JP, Vandamme EJ, Vlerick C (1973) Z Allg Mikrobiol 13: 355
179. Gröger G, Schmauder HP, Mothes K (1966) Z Allg Mikrobiol 6: 275
180. Biselli M, Kragl U, Wandrey C (1995) Reaction engineering for enzyme catalyzed biotransformations. In: Drauz K, Waldheim H (eds) Enzymes in organic synthesis. VCH, Weinheim, p 89
181. Nicholls, A (1992) Dept. of Biochemistry and Molecular Biophysics, Columbia University, New York
182. Barburina I, Moore DJ, Volkov A, Kahyaoglu A, Jordon F, Mendelsohn R (1996) Biochemistry 35: 10249

Screening of Novel Microbial Enzymes for the Production of Biologically and Chemically Useful Compounds

Sakayu Shimizu, Jun Ogawa, Michihiko Kataoka and Michihiko Kobayashi
Department of Agricultural Chemistry, Kyoto University,
Kitashirakawa-oiwakecho, Sakyo-ku, Kyoto 606, Japan

Dedicated to Professor Dr. Maria-Regina Kula on the occasion of her 60th birthday

Enzymes have been generally accepted as superior catalysts in organic synthesis. Micro-organisms in particular have been regarded as treasure sources of useful enzymes. The synthetic technology using microbial enzymes or micro-organisms themselves is called microbial transformation. In designing a microbial transformation process, one of the most important points is to find a suitable enzyme for the reaction of interest. Various kinds of novel enzymes for specific transformations have been discovered in micro-organisms and their potential characteristics revealed. This article reviews our current results on the discovery of novel enzymes for the production of biologically and chemically useful compounds, and emphasizes the importance of screening enzymes in a diverse microbial world.

Advances in Biochemical Engineering/
Biotechnology, Vol. 58
Managing Editor: Th. Scheper
© Springer-Verlag Berlin Heidelberg 1997

1 Introduction

During the last twenty years, biochemical reactions performed by micro-organisms or catalyzed by microbial enzymes have been extensively evaluated from the viewpoint of synthetic organic chemistry, and as a consequence they have been shown to have a high potential for both theoretical and practical applications in synthetic chemistry. Many attempts to utilize biological reactions for practical synthetic processes have been made – for example, for the preparation of pharmaceuticals, fine chemicals, food additives, and commodity chemicals. Such synthetic technology is called microbial transformation, or alternatively, microbial conversion, biotransformation, bioconversion, or enzymation [1, 2].

A microbial transformation is the conversion of one substance (substrate) to another (product) by a micro-organism. It is a chemical reaction, catalyzed by a particular cellular enzyme or by an enzyme originally produced within cells. Most such enzymes are necessary for the normal functioning of the biological processes of cellular metabolism and reproduction. In microbial transformation, however, these enzymes simply act as catalysts (biocatalysts) for chemical reactions. In addition to their natural substrates, many of these enzymes can utilize other structurally related compounds as substrates and therefore occasionally catalyze unnatural reactions when foreign substrates are added to the reaction medium. Thus, microbial transformation constitutes a specific category of chemical synthesis.

In designing a microbial conversion process many important aspects require careful consideration, for example, the selection of a compound to be synthesized, a survey of available substrates, and the routes or reactions needed. Another point (the most important one) is to find microbial enzymes which are suitable for the processes designed, and to subsequently evaluate the enzyme's potential. Moreover, the discovery of a new enzyme or a new reaction provides a clue for designing a new microbial transformation process. To find microbial enzymes that are suitable as potent catalysts, the capabilities of well-known enzymes or reactions need to be reassessed, and novel microbial strains or enzymes need to be discovered. Screening may be one of the most efficient and successful ways of searching for new or suitable microbial enzymes.

Screening should be carried out for a wide variety of micro-organisms because:

(1) Enzymes are as good as chemical catalyst for catalyzing a variety of chemical reactions. We already know of more than 3000 reactions catalyzed by enzymes [3]; some of them were even more effective than corresponding chemical catalysts. (2) There is no limit to the number of micro-organisms that can be tested in nature and they are all quite different from each other. They modify or degrade a great variety of complex organic compounds, and so at least one of them can be expected to catalyze a given reaction.

In our laboratory, by means of extensive screening, many novel microbial enzymes, which have great potential for the practical production of biologically and chemically useful compounds, were found and their superior properties were revealed from biochemical and practical viewpoints. In this review, we shall introduce some successful examples and describe the valuable properties of novel microbial enzymes.

2 Hydantoinases and Decarbamoylases

2.1 D-Amino Acid Production by D-Hydantoinase and D-Decarbamoylase

D-p-Hydroxyphenylglycine and its derivatives are important as side-chain precursors for semisynthetic penicillins and cepharosporines. Yamada and co-workers of our laboratory found that these amino acids can be efficiently prepared from the corresponding 5-monosubstituted hydantoins using the microbial enzyme D-hydantoinase [4].

The bacterial D-hydantoinase has been isolated as crystals from cells of *Pseudomonas putida* (= *P. striata*) (Table 1) [5]. Because the purified enzyme showed the highest activity and affinity toward dihydrouracil, the enzyme was identified as dihydropyrimidinase (EC. 3.5.2.2). Interestingly, the enzyme also attacked a variety of aliphatic and aromatic D-5-monosubstituted hydantoins, yielding the corresponding D-form of N-carbamoyl-α-amino acids. Thus, the enzyme can be used for the preparation of various D-amino acids. Under the conditions used for the enzymatic hydrolysis of hydantoin at pH 8 to 10, the L-isomers of the remaining hydantoins are racemized through base catalysis. Therefore, the racemic hydantoins can be converted quantitatively into N-carbamoyl-D-amino acids through this step.

Decarbamoylation to D-amino acid was performed by treating the N-carbamoyl-D-amino acid with equimolar nitrite under acidic conditions [6]. But now, this step can also be carried out enzymatically. Recently, Shimizu and co-workers found a novel enzyme, D-decarbamoylase (N-carbamoyl-D-amino acid amidohydrolase), which stereospecifically hydrolyzes N-carbamoyl-D-amino acids, in several bacteria [7, 8]. For example, *Blastobacter* sp. A17p-4 was found to produce D-decarbamoylase together with D-hydantoinase [8]. Therefore, a sequence of two enzyme-catalyzed reactions, the D-stereospecific hydrolysis of DL-5-(p-hydroxyphenyl) hydantoin and subsequent hydrolysis of the D-carbamoyl derivative to D-p-hydroxyphenylglycine, is possible (Fig. 1). Based on these results, a new commercial process for the production of D-p-hydroxyphenylglycine has been developed [9].

Table 1. Properties of cyclic ureide-hydrolyzing enzymes

Properties	D-Hydantoinase		Imidase	N-Methylhydantoin amidohydrolase
	Pseudomonas putida IFO 12996	*Blastobacter* sp. A17p-4	*Blastobacter* sp.A17p-4	*P. putida* 77
Native M_r	190,000	200,000	105,000	290,000
Subunit M_r (SDS-PAGE)	48,000	53,000	35,000	70,000 & 80,000
Number of subunit	4	4	3	$\alpha_2\beta_2$
Substrate specificity (stereospecificity)	dihydropyrimidines 5-monosubstituted hydantoins (D)	dihydropyrimidines 5-monosubstituted hydantoins (D)	cyclic imides dihydropyrimidines	*N*-methylhydantoin 5-monosubstituted hydantoins (L)
Optimum pH				
ring-opening hydrolysis	8.0–9.0	9.0–10.0	7.5–8.0	8.0
cyclyzing dehydration	n.d.	5.0	6.5	(not catalyze)
Optimum temperature	55°C	60°C	60°C	37°C
pH stability	6.0–7.0	5.0–8.5	6.0–9.0	5.5–8.5
Thermal stability	< 60°C	< 60°C	< 60°C	< 40°C
Metal ion requirement	stabilization Co²⁺	activation Ni²⁺, Co²⁺, Mn²⁺	Co²⁺	Mg²⁺ and K⁺
Inhibitor	SH-inhibitors metal-ion chelators	SH-inhibitors Hg²⁺	SH-inhibitors heavy metal ions metal-ion chelators diisopropylphosphofluoridate	SH-inhibitors heavy metal ions metal-ion chelators carbonyl inhibitors

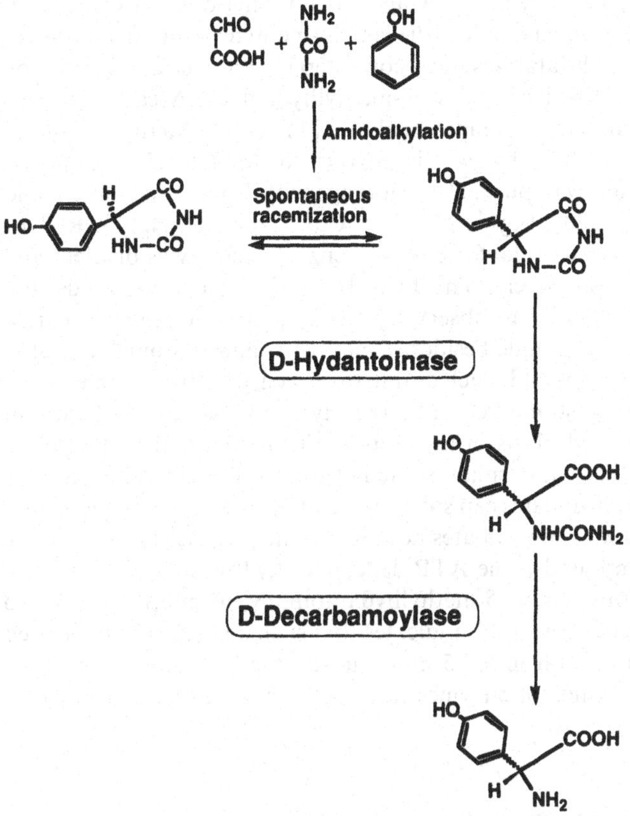

Fig. 1. Chemical and enzymatic process for the production of D-*p*-hydroxyphenylglycine from DL-5-(*p*-hydroxyphenyl) hydantoin

2.2 Diversity of Hydantoinases

Many kinds of enzymes with different substrate specificities are involved in hydantoin hydrolysis. Ogawa et al. [10] found two hydantoin-hydrolyzing enzymes in *Blastobacter* sp. A17p-4. These enzymes were purified to homogeneity and characterized (Table 1). One hydrolyzed dihydropyrimidines and 5-monosubstituted hydantoins to the corresponding *N*-carbamoyl amino acids. Since the hydrolysis of 5-substituted hydantoins by this enzyme was D-stereospecific, this enzyme was identified as D-hydantoinase, which is identical with dihydropyrimidinase. The other one preferably hydrolyzed cyclic imide compounds such as glutarimide and succinimide more than cyclic ureide compounds such as dihydrouracil and hydantoin. Because there have been no reports on enzymes which show same substrate specificity as this enzyme, it is considered to be a novel enzyme, which should be called imidase [10].

Other than these, an enzyme showing L-hydantoinase activity was also found. It is an ATP-dependent amidohydrolase, N-methylhydantoin amidohydrolase (EC 3.5.2.14), which catalyzes the second step reaction in the degradation route from creatinine to glycine, via N-methylhydantoin, N-carbamoylsarcosine, and sarcosine as successive intermediates [11–19]. N-Methylhydantoin amidohydrolase requires ATP hydrolysis to ADP for the hydrolysis of amide substrates. This enzyme was purified to homogeneity from P. putida 77 and characterized (Table 1) [20]. Mg^{2+}, Mn^{2+} or Co^{2+}, and K^+, $NH4^+$, Rb^+ or Cs^+ were concomitantly essential for ensuring enzyme activity as divalent and monovalent cations, respectively. The hydrolysis of amide compounds and coupled hydrolysis of ATP were observed with hydantoin, L-5-methylhydantoin, glutarimide and succinimide besides N-methylhydantoin. Some naturally-occurring pyrimidine compounds such as dihydrouracil, dihydrothymine, uracil and thymine, effectively stimulate ATP hydrolysis by the enzyme without undergoing detectable hydrolysis of themselves. Furthermore, the enzyme exhibits unique nucleoside triphosphatase activity [21]. Besides ATP, various kinds of nucleoside triphosphates can substitute and be hydrolyzed to yield their corresponding nucleoside diphosphates not only in the presence but also in the absence of amide compounds. The ATP-dependent hydrolysis of 5-monosubstituted hydantoins, for example 5-methylhydantoin, by the enzyme proved to be L-isomer specific [20]. This result indicates that such an enzyme can be used for producing L-amino acid from DL-5-monosubstituted hydantoin.

Above all, various kinds of enzymes have been found to be concerned in hydantoin hydrolysis (Fig. 2).

2.3 Diversity of Decarbamoylases

Since D-hydantoinase was identified as dihydropyrimidinase, it is proposed that D-amino acid production from DL-5-monosubstituted hydantoins involves the action of the series of enzymes involved in the pyrimidine degradation pathway. Based on this proposal, D-decarbamoylase was thought to be identical with β-ureidopropionase (EC 3.5.1.6) which functions in pyrimidine metabolism.

To confirm this proposition, D-decarbamoylases (N-carbamoyl-D-amino acid amidohydrolase) were first purified homogeneously from Comamonas sp. E222c [7] and Blastobacter sp. A17p-4 [8], and characterized (Table 2). N-Carbamoyl-D-amino acids having hydrophobic groups served as good substrates for these enzymes. Since these enzymes strictly recognize the configuration of the substrate and only hydrolyze the D-isomer of the N-carbamoyl amino acid, they can be used for the resolution of racemic N-carbamoyl amino acid. These enzymes did not hydrolyze β-ureidopropionate, suggesting that these enzymes are different from β-ureidopropionase.

To make clear the difference between D-decarbamoylase and β-ureidopropionase, the β-ureidopropionase from aerobic bacteria was purified and characterized for the first time [22]. The specific features of the purified enzyme

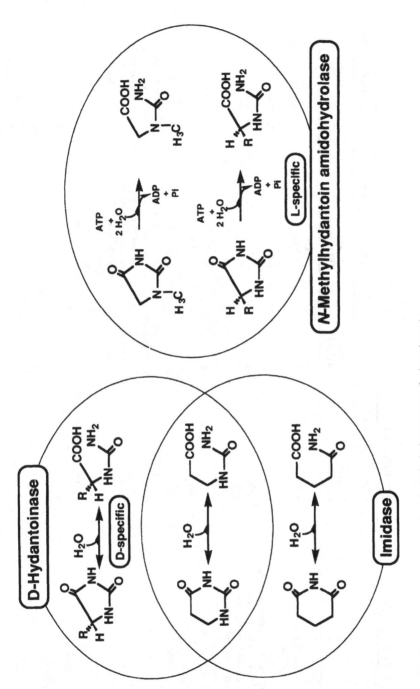

Fig. 2. Reactions catalyzed by hydantoin-hydrolyzing enzymes (hydantoinases)

Table 2. Properties of N-carbamoyl amino acid-hydrolyzing enzymes

Properties	D-Decarbamoylase		β-Ureidopropionase	L-Decarbamoylase
	Comamonas sp. E222c	Blastobacter sp. A17p-4	P. putida IFO12996	Alcaligenes xylosoxidans
Native M_r	117,000	111,000	95,000	134,000
Subunit M_r (SDS-PAGE)	40,000	40,000	44,000	65,000
Number of subunit	3	3	2	2
pI	4.0	5.3	n.d.	n.d.
Substrate specificity (stereospecificity)				
N-carbamoyl-α-amino acid	O(D)	O(D)	O(L)	O(L)
N-carbamoyl-β-amino acid	–	–	O	–
N-carbamoyl-γ-amino acid	–	–	O	–
Optimum pH	8.0–9.0	8.0–9.0	7.5–8.2	8.0–8.3
Optimum temperature	40°C	55°C	60°C	35°C
pH stability	7.0–9.0	6.0–9.0	6.0–8.5	6.0–9.5
Thermal stability	<40°C	<50°C	<65°C	<30°C
Metal ion requirement	none	none	Co^{2+}, Ni^{2+}, Fe^{2+}, Mn^{2+}	Co^{2+}, Ni^{2+}, Mn^{2+}
Inhibitor	SH-inhibitors heavy metal ions	SH-inhibitors heavy metal ions N-carbamoyl-L-amino acids ammonia	SH-inhibitors heavy metal ions metal-ion chelators carbonyl inhibitors N-carbamoyl-D-amino acids sodium propionate	SH-inhibitors heavy metal ions metal-ion chelators carbonyl inhibitors sodium propionate

n.d.: not determined
O: hydrolyzed
–: not hydrolyzed

from *P. putida* IFO 12996 (Table 2) are quite different from those of the β-ureidopropionases from other sources, such as mammals [23, 24], protozoa [25] and anaerobic bacteria [26], and D-decarbamoylase in structure, in metal-ion dependency and especially in substrate specificity. The enzyme showed a broad substrate specificity; not only *N*-carbamoyl-β-amino acids, but also *N*-carbamoyl-γ-amino acids, and several *N*-carbamoyl-α-amino acids such as *N*-carbamoylglycine, *N*-carbamoyl-L-alanine, *N*-carbamoyl-L-serine, and *N*-carbamoyl-L-α-amino-*n*-butyrate are hydrolyzed. The hydrolysis of *N*-carbamoyl-α-amino acids is strictly L-enantiomer specific, i.e., the reverse stereospecificity to those of D-decarbamoylase.

N-Carbamoyl-L-amino acid-hydrolyzing activities have been found in micro-organisms [27–31], and suggested to be useful for the L-amino acid production. To reveal whether these activities derived from β-ureidopropionase activity, the enzyme, L-decarbamoylase (*N*-carbamoyl-L-amino acid amidohydrolase) which hydrolyzes *N*-carbamoyl-L-amino acid, was purified from *Alcaligenes xylosoxidans* and characterized (Table 2) [32]. The enzyme from *A. xylosoxidans* resembles β-ureidopropionase in structure and in metal-ion dependency. However, the substrate specificity of the enzyme was different from that of β-ureidopropionase. β-Ureidopropionase from *P. putida* IFO 12996 hydrolyzed short chain *N*-carbamoyl-L-amino acids, but not long chain aliphatic and aromatic *N*-carbamoyl-L-amino acids [22]. On the other hand, L-decarbamoylase showed a broad substrate specificity not only for short chain but also for long chain and aromatic *N*-carbamoyl-L-amino acids. Furthermore, the enzyme did not hydrolyze β-ureidopropionate; suggesting that the enzyme is distinct from β-ureidopropionase in function. L-Decarbamoylase can be used for the optical resolution of racemic *N*-carbamoyl amino acids.

Above all in the hydrolysis of *N*-carbamoyl amino acid, three types of enzymes with different substrate specificity function (Fig. 3).

2.4 Application of Hydantoinases and Decarbamoylases

Hydantoinases and decarbamoylases have been applied for the production of optically active amino acids from DL-5-monosubstituted hydantoins. A variety of enzymes have been reported elsewhere. Runser et al. [33] reported the occurrence of D-hydantoinase without dihydropyrimidinase activity. Watabe et al. [34] reported that an ATP-dependent hydantoin-hydrolyzing enzyme is involved in the L-amino acid production from DL-5-monosubstituted hydantoin by *Pseudomonas* sp. NS671. This enzyme shows no stereospecificity. Hydantoinase showing no stereospecificity and not requiring ATP was also reported [35]. Recently, hydantoin-racemizing enzymes were found [36, 37]. These enzymes make it possible to totally convert racemic substrates, which only slowly racemize under reaction conditions, to a single stereoisomer. The combinations of these hydantoin-transforming enzymes provide a variety of processes for optically active amino acid production (Fig. 4).

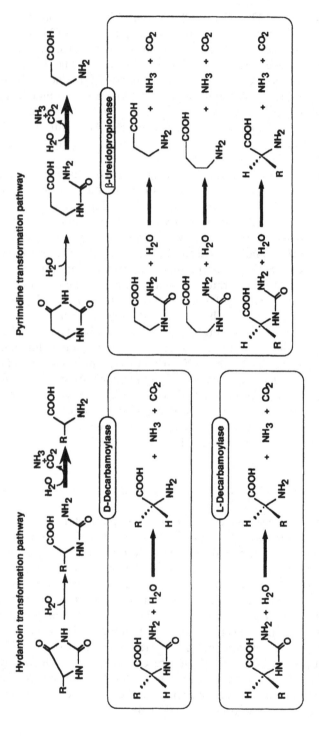

Fig. 3. Reactions catalyzed by *N*-carbamoyl amino acid-hydrolyzing enzymes (decarbamoylases)

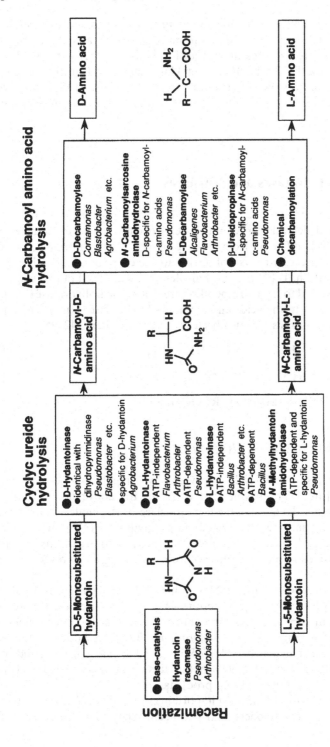

Fig. 4. Processes for the production of optically active amino acids with hydantoinases and decarbamoylases

Apart from this, N-methylhydantoin amidohydrolase is also useful for diagnostic purposes. N-Methylhydantoin amidohydrolase and the corresponding enzyme in microbial creatinine transformation [i.e. N-carbamoylsarcosine amidohydrolase (EC 3.5.1.59)] are useful for enzymatically measuring creatinine in serum and urine, which is a marker of renal dysfunction [12]. This enzymatic method has proved to be simple and precise, and shows excellent sensitivity and specificity [38].

3 Nitrile Hydratases and Nitrilases

The microbial degradation of nitriles can occur via two different enzymatic pathways [39], different from cyanogenesis [40]: (i) nitrilase (EC 3.5.5.1) catalyzes the direct hydrolysis of nitriles to the corresponding carboxylic acids and ammonia (Eq. 1) [41]; and (ii) nitriles are catabolized in two stages – they are first converted to the corresponding amides by nitrile hydratase (EC 4.2.1.84) (Eq. 2), and then to the acids and ammonia by amidase (Eq. 3). The microbial degradation of nitriles has also been reviewed elsewhere [42–45].

$$RCN + 2H_2O \rightarrow RCOOH + NH_3 \qquad\qquad [Eq.\ 1]$$

$$RCN + H_2O \rightarrow RCONH_2 \qquad\qquad [Eq.\ 2]$$

$$RCONH_2 + H_2O \rightarrow RCOOH + NH_3 \qquad\qquad [Eq.\ 3]$$

3.1 Diversity of Nitrile Hydratases

Nitrile hydratase was initially discovered by Yamada and co-workers of our laboratory [46]. The enzyme that catalyzes the conversion of nitriles into amides is clearly distinguishable from nitrilase in the mode of nitrile-degradation. Thus, the enzyme was termed "nitrile hydratase (NHase)". This enzyme was purified and characterized from various micro-organisms. NHases are roughly classified into two groups according to the metal involved: Fe-type and Co-type [45].

Pseudomonas chlororaphis B23 with high NHase activity was isolated as an isobutyronitrile-assimilating bacterium [47]. The NHase acts well on acrylonitrile to form acrylamide but the amidase hardly acts on acrylamide in this strain. When resting cells of P. chlororaphis B23 were added to the reaction mixture and incubated at 10 °C for 7.5 h, more than 400 g of acrylamide per liter was accumulated. More than 99% of the substrate was converted to acrylamide without the formation of any by-products.

Acrylamide is one of the most important commodity chemicals and is used in coagulators, soil conditioners, and petroleum recovering agents. Conventional

Cu-catalytic process

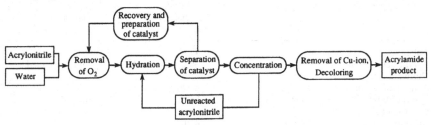

Microbial process

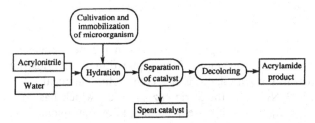

Fig. 5. Comparative flowsheet for the conventional and microbial processes for the production of acrylamide

chemical synthesis involves the hydration of acrylonitrile with the use of copper salts as a catalyst (Fig. 5). However, this chemical method has various problems: (i) the rate of acrylic acid formation is higher than that of acrylamide formation, (ii) the double bond of both substrate and product causes the formation of by-products such as nitrylotrispropionamide and ethylene cyanohydrin, and (iii) polymerization occurs at the double bond of both substrate and product. The biotransformation of acrylonitrile to acrylamide by the *P. chlororaphis* B23 NHase (induced by methacrylamide [48]) overcame the above problems. However, there was a problem that when *P. chlororaphis* B23 is grown on the medium containing sucrose, mucilaginous polysaccharides, which resemble levan, are produced [49]; it is not easy to harvest the cells by brief centrifugation due to the high viscosity of the culture medium. This is a serious technical disadvantage that cannot be overlooked. The accumulation of these polysaccharides also has a bad effect on aeration during the cultivation, resulting in a waste of the energy derived from sucrose as one of carbon sources. Therefore, mucilage polysaccharide-nonproducing mutants were isolated by the chemical mutagenesis method using *N*-methyl-*N'*-nitro-*N*-nitrosoguanidine. The resultant mutant, which can be easily precipitated after brief centrifugation and which has high NHase activity, was bred from the parent strain. Consequently, the process for the production of acrylamide from acrylonitrile by the *P. chlororaphis* B23 NHase was established on an industrial scale (Table 3). This

Table 3. Improvement of biocatalyst for the acrylamide production

	Pseudomonas chlororaphis B23	Rhodococcus rhodochrous J1
Tolerance to acrylamide (%)	40	50
Acrylic acid formation	barely detected	barely detected
Cultivation time (h)	45	72
Activity of culture (units/ml)	1,400	2,100
Specific activity (units/mg cells)	85	76
Cell yield (g/l)	17	28
Acrylamide productivity (g/g cells)	850	> 7, 000
Total amount of production (ton/year)	6,000	> 30, 000
Final concentration of acrylamide (%)	27	40
First year of production scale	1988	1991

was the first case in which biotechnology was applied in the petrochemical industry and also the first successful example of the application of an industrial biotransformation process in the manufacture of a commodity chemical.

The *P. chlororaphis* B23 NHase is the ferric enzyme [50], which has been characterized in detail [51, 52]. (i) The NHase is the first known nonheme iron enzyme containing a typical low-spin Fe(III) site, (ii) the axial position of the Fe(III) site in the enzyme may be occupied by aquo and sulfhydryl groups, and (iii) aliphatic nitrile substrates directly bind to the Fe(III)-active center through H_2O-substrate replacement. The NHase also seems to contain pyrroloquinoline quinone (PQQ) or a PQQ-like compound [53].

A ferric NHase from *Rhodococcus* sp. N-771 [54], which is probably the same as that from *Rhodococcus* sp. N-774 because of the coincidence in amino acid sequences, shows a photosensitive phenomenon both in vivo and in vitro [55, 56]. Its NHase activity is increased by exposure to near ultraviolet light. Absorption and fluorescence spectra indicated that the chromophore involved in the photoactivation is the iron complex. Crystallographic parameters of the *Rhodococcus* sp. N-771 NHase have been determined by precision X-ray diffraction studies [57].

Another type of metal-containing NHase was found in *Rhodococcus rhodochrous* J1. This strain produces two kinds of NHases [39]: a high-molecular-mass (520 kDa) NHase (H-NHase) [58, 59] and a low-molecular-mass (130 kDa) NHase (L-NHase) [60]. When *R. rhodochrous* J1 is cultured in a medium containing urea and cyclohexanecarboxamide in the presence of cobalt ions, H-NHase and L-NHase are selectively induced, respectively, while both NHases are induced by culture in the presence of crotonamide and cobalt ions. Both H- and L-NHases contain cobalt ions as a co-factor in contrast to the iron-ion-containing NHases from *P. chlororaphis* B23, *Rhodococcus* sp. N-774 and *Brevibacterium* R312. H- and L-NHases differ in their physicochemical properties and substrate specificities: H-NHase acts preferentially on aliphatic nitriles, while L-NHase has a higher affinity than H-NHase for aromatic nitriles [39].

There is a new type of NHase containing both cobalt and ferric ions in *Agrobacterium tumefaciens* [61, 62]. This NHase is involved in the biosynthesis of a plant hormone: indole-3-acetic acid from indole-3-acetonitrile. This novel biosynthetic pathway was discovered in the family Rhizobiaceae containing two genera, *Agrobacterium* and *Rhizobium*, although the latter species is a beneficial plant symbiont that induces nitrogen-fixing nodules on the roots of legumes while the former species is a phytopathogen. The purified *A. tumefaciens* NHase catalyzes stoichiometrically the hydration of indole-3-acetonitrile into indole-3-acetamide with a lower *Km* [61].

3.2 Organization and Regulation of Nitrile Hydratase Genes

Various NHase genes have been cloned and characterized [39] (Fig. 6): from *R. rhodochrous* J1 [63] and *P. chlororaphis* B23 [64], from *Rhodococcus* sp. N-774 [65, 66] and *Rhodococcus erythropolis* [67], from *Brevibacterium* R312 [68] and *Rhodococcus* sp. [69]. Gene constructions of α- and β-subunits of each NHase are variously organized in the order [39]. An amidase gene is present just upstream from each NHase gene in *P. chlororaphis* B23, *Rhodococcus* sp. N-774, *R. erythropolis*, *Brevibacterium* R312, and *Rhodococcus* sp., while an amidase gene [70] is 1.9-kb downstream of the L-NHase gene in *R. rhodochrous* J1. The close location of NHase and amidase genes strongly supports the theory that both enzymes are closely related in the two-step reaction of the nitrile-degradation pathway. In *P. chlororaphis* B23, two open reading frames, P47K and OrfE, are present just downstream of the coding region of the β-subunit of the NHase. The close arrangement of five genes (amidase, α- and β-subunits of the NHase, P47K, and OrfE, in this order) suggests that these proteins seem to be translated from a single mRNA species. An additional 38-kDa protein, probably encoded by the region upstream of the *P. chlororaphis* B23 amidase gene, is also produced in large amounts in the *Escherichia coli* transformant, while amidase is not, as measured by SDS-polyacrylamide gel electrophoresis. The close association of the NHase and amidase genes simplifies achieving NHase expression in *E. coli*. P38K as well as P47K and OrfE also seems to be important for efficient expression of NHase activity in the transformant.

Both H- and L-NHase genes are expressed under the control of a *lac* promoter in *E. coli* only when the transformants are cultured in the presence of CoCl$_2$ [63]. However, the level of NHase activity in their cell-free extracts is much lower than those of H- and L-NHases in *R. rhodochrous* J1. Most of H-NHase is produced as an insoluble form in the *E. coli* transformant, and there is only a little L-NHase in either the supernatants or precipitates of the extracts. Establishment of an effective host-vector system in *Rhodococcus* [71] facilitates the development of strains with improved NHase activities. In this connection, the transformation system in *Rhodococcus* has been investigated. Enzyme assays of recombinant *Rhodococcus* cells showed that a downstream region of the *Rhodococcus* sp. N-774 NHase gene is indispensable for the production of active

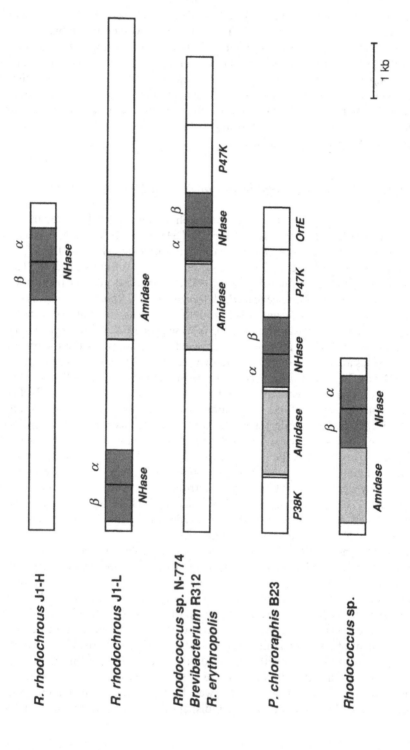

Fig. 6. Gene organization of various NHases and amidases

NHase [56]. The regulation mechanism of both H- [59] and L-NHase [60] genes have also been characterized using this system.

3.3 Transformation of Nitriles to Amides by Nitrile Hydratase

When *R. rhodochrous* J1 is cultured in the optimum culture medium containing urea, glucose, peptone, yeast extract, K_2HPO_4, KH_2PO_4, and $CoCl_2$, the amount of H-NHase in the cell-free extracts accounts for more than 50% of the total soluble protein [39]. This purified NHase acts well on acrylonitrile. Together with these characteristics, the following advantages for the acrylamide production are observed in H-NHase, leading to the establishment of the industrial manufacture by this NHase in place of the *P. chlororaphis* B23 NHase. Urea as a potent inducer for the NHase formation is much cheaper than methacrylamide, which had been used as an inducer for the *P. chlororaphis* B23 NHase. The *R. rhodochrous* J1 H-NHase is much more heat stable and more resistant to a high concentration of acrylonitrile than the *P. chlororaphis* B23 NHase. The former enzyme is also peculiar in its extremely high tolerance to acrylamide. The hydration reaction of acrylonitrile catalyzed by the H-NHase-containing cells proceeds even in the presence of 50% (w/v) acrylamide [72]. Despite the use of a different biocatalyst, almost no change was required in the manufacturing plant (production capacity: 30,000 tons/year) (Table 3).

The *R. rhodochrous* J1 NHases act not only on acrylonitrile but also on other aliphatic nitriles. Furthermore, aromatic nitriles are also good substrates for the enzyme. Therefore, exchange of acrylonitrile as a substrate with other nitriles enables us to synthesize various useful amides, a good example being nicotinamide [73], a vitamin widely used in animal feed supplementation. The *R. rhodochrous* J1 cells containing both H- and L-NHases, which are induced by the addition of crotonamide to the culture medium, have high activity toward 3-cyanopyridine. This manufacture is scheduled to begin 1997. Other applications of NHase for the production of useful amides and the stereo- and regio-specific-conversion of nitriles are now being increasingly recognized.

3.4 Bacterial Nitrilases

Nitrilase was initially discovered in plants as an enzyme involved in the bio-synthesis of the plant hormone indole-3-acetic acid (IAA) [74, 75]. Recently, four genes of nitrilases (belonging to arylacetonitrilase) involved in the IAA biosynthesis have been cloned and characterized from *Arabidopsis thaliana* [76–78]. After the discovery of the plant nitrilase in 1964, various nitrilases were purified and characterized [41]. Nitrilases are roughly classified into three major categories according to substrate specificity: (i) aromatic nitrilase, which acts on aromatic or heterocyclic nitriles; (ii) aliphatic nitrilase, which acts on aliphatic nitriles: (iii) arylacetonitrilase, which acts on arylacetonitriles. These three types of nitrilases have been studied at both protein and gene levels by using several kinds of nitrilase-producing bacteria. Nitrilases previously reported were very labile and were not obtained in amounts large enough for characterization. *R. rhodochrous* J1 was found to over produce the aromatic nitrilase inducibly by the addition of nitriles to a medium [79]; particularly, the amount of the nitrilase overproduced by isovaleronitrile is 35% of all soluble protein. Heterocyclic nitriles as well as aromatic nitriles are attacked by the purified enzyme [80]. The aliphatic nitrilase [81] is also strongly induced by isovaleronitrile in *R. rhodochrous* K22 [82]; the nitrilase formed is 24% of all soluble protein.

All nitrilases belonging to aromatic nitrilase and arylacetonitrilase are susceptible to SH-reagents, except the *R. rhodochrous* K22 aliphatic nitrilase [81]. Analysis by site-directed mutagenesis for our three nitrilases from *R. rhodochrous* J1 [83], *R. rhodochrous* K22 [84] and *Alcaligenes faecalis* JM3 [85] revealed that a unique cysteine that is conserved at the corresponding position in each nitrilase is essential for the catalytic activity (Fig. 7). All nitrilases whose structural genes are cloned have a similar amino acid sequence; nitrilase forms a superfamily.

3.5 Applications of Nitrilases

Applications of these nitrilases are now being increasingly recognized [78]: for example, for the production of useful acids, and for the stereo- and regio-specific conversion of nitriles. Using the nitrilase overproduced in the *R. rhodochrous* J1 cells, the production of vitamins such as nicotinic acid (172 g/l) from 3-cyanopyridine [86], and *p*-aminobenzoic acid (110 g/l) from *p*-aminobenzonitrile [87] have been reported. An antimycobacterial agent, pyrazinoic acid (434 g/l) is produced from cyanopyrazine by the *R. rhodochrous* J1 cells containing the aromatic nitrilase [88]. The *R. rhodochrous* J1 nitrilase selectively converts aromatic dinitriles such as isophthalonitrile and terephthalonitrile into 3-cyanobenzoic acid and 4-cyanobenzoic acid, respectively [89]. The *R. rhodochrous* K22 nitrilase is also capable of hydrolyzing only one cyano group of a dinitrile, glutaronitrile, to a carboxyl group [90]. Similar specificities were reported in the reaction by the *R. rhodochrous* NCBI11216 nitrilase for 1,3-dicyanobenzene [91] and fumaronitrile [92] as a dinitrile substrate.

Transgenic plants containing a nitrilase specific for the herbicide bromoxynil (= 3,5-dibromo-4-hydroxybenzonitrile) have also been developed [93]; the Calgene company transformed tobacco plants with the bacterial *Klebsiella ozaenae* gene encoding nitrilase [94] that detoxifies the herbicide by hydrolysis (conversion of bromoxynil to 3,5-dibromo-4-hydroxybenzoic acid), resulting in the establishment of the herbicide-resistant transgenic plants.

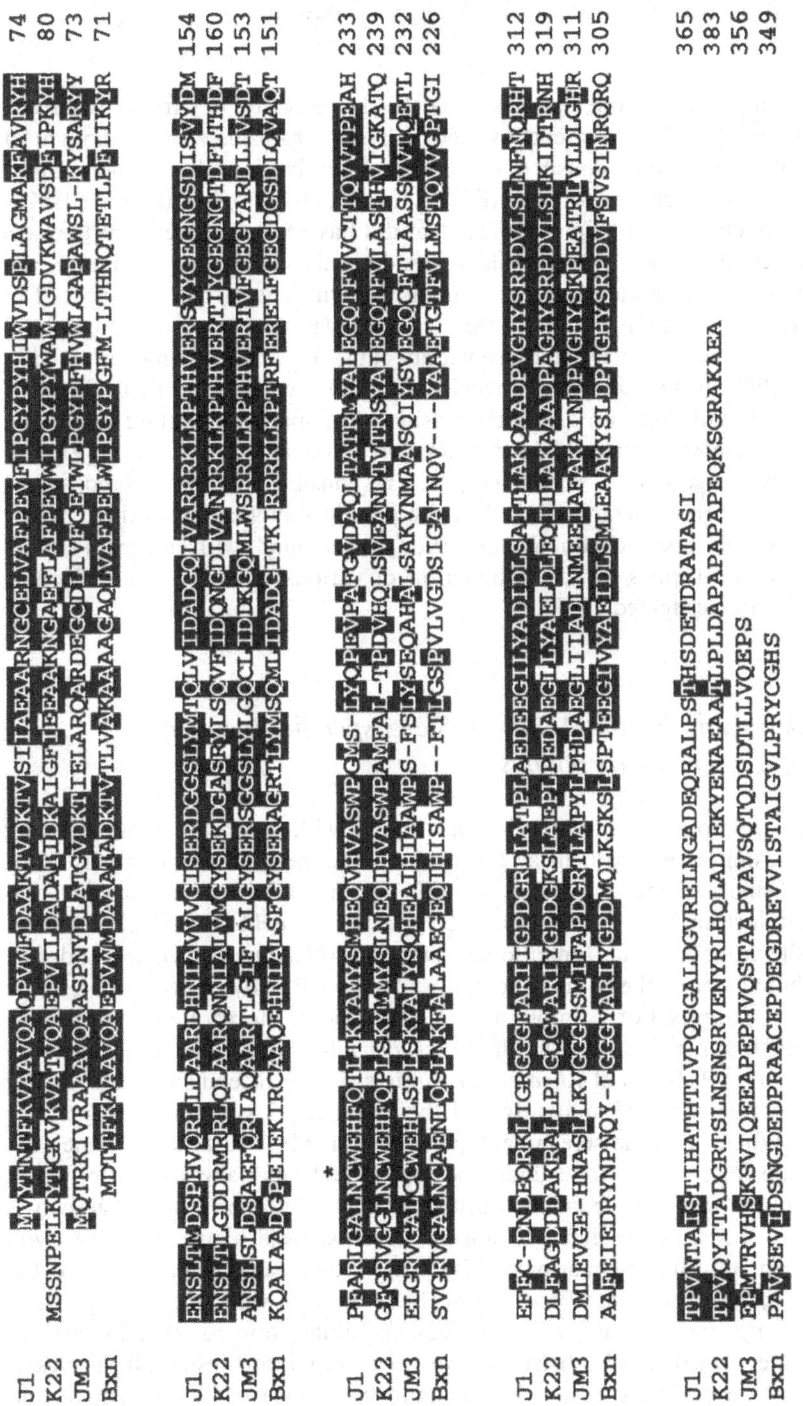

```
J1    MVYTNTFKVAAVQAQPVWLDAAKTVDKTVSITAEAARNGCELVAFPEVFIPGYPYHIWVDSPLAGMAKFAVRYH      74
K22   MSSNPELKYTGKVKVATVQAEPVIIDADATIDKAIGFIEEAAKNGAEFTAFPEVWIPGYPYWAIGDVKWAVSDFIPKYH  80
JM3   MQTRKIVRAAVQAASPNYDIATGVDKTIELARQARDEGCDIVFGETWIPGYPFEVWLGAPAWSL-KYSARYY          73
Bxn   MDTYFTFKAAAVQAEPVWMDAAATADKTVTLVAKAAAGAQLVAFPEIWIPGYPGFM-LTHNQTETLPFIIKYR         71

                                              *
J1    ENSLTMDSPHVQRLIDAARDHNIAVVGISERDGGSLYMQLVIDADGQIVARRRKLKPTHVERSVYGEGNGSDISVYDM   154
K22   ENSLTRLGDDRMRRLQIAARONNIAIVMGYSEKDGASRYLSQVVFIDOYGDIVANRRKLKPTHVERTIYGEGNGTDFLTHDF 160
JM3   ANSLSLDSAEFQRIAQAARTLGIFTAIGYSERSGGSLYLGQCIIDEKGQMLWSRRKLKPTHVERIVFGGEYARDHIVSDT   153
Bxn   KQATAADGPEIEKIRCAAQEHNIAISFGYSERAGRTLYMSQMTIDADGITKIRRRKLKPTRFERELLGEGDGSDLQVAQT   151

J1    PFARIGAINCWEHFQITTKYAMYSMHEQVHVASWPGMS--LYQPEVPAFGVDAQLTATRMYALEGOTFVCVTTQVVTPEAH  233
K22   GFGRVGGINCWEHFQPISKYLMYSLNEQIHVASWPAMFAT-TPDVHQLSVEANDIVTERSYAIEGQTFVLASTHVIGKATQ  239
JM3   ELGRVGAICCWEHLSPISKYATYSQHEAIHIAAWPS-FSLYSEQAHTLSAKVNMAASQIYSVEGQCFTLAASSVVTQETL   232
Bxn   SVGRVGAINCAENLQSINKFALAAEGEQIHISAWP---FTLGSPVLVGDSIGAINQV---YAAEIGTFVLMSTQVVGPTGI   226

J1    EFFC-DNDEQRKIIGRGGGFARIIGPDGRDLAITPLAEDEEGILYADIDLSAITLAKOAADPVGHYSRPDVLSLNFNORHT  312
K22   DIFAGDDDAKRAIIPIGQGWARIYGPDGKSIAEPIPEDAEGILYAETIDLFQIHLAKAADPAGHYSRPDVLSLKIDTRNH   319
JM3   DMLEVGE-HNASILKVGCGSSMIFAPDGRILAPYTPHDAEGLIIADLNMEEIAFAKAINDPVGHYSKPEATRIVLDLGHR   311
Bxn   AAFEIEDRYNPNQY-LGGGYARIYGPDMQLKSKSLSPTEEGIVARIDLSMLEAAKYSLDPTGHYSRPDVFSVSINRQRQ   305

J1    TPVNTAISTIHATHTLVPQSGALDGVRELNGADEQRALPSTHSDETDRATASI                             365
K22   TPVQYITTADGRTSLNSNSRVENYRLHQLADIEKYENAEAATLPLDAPAPAPEQKSGRAKAEA                    383
JM3   EPMTRVHSKSVIQEEAPEHVQSTAAPVAVSQTQDSDTLLVQEPS                                       356
Bxn   PAVSEVIDSNGDEDPRAACEPDEGDREVVISTAIGVLPRYCGHS                                       349
```

Fig. 7. Alignment of amino acid sequences in bacterial nitrilases: J1, *R. rhodochrous* J1; K22, *R. rhodochrous* K22; JM3, *A. faecalis* JM3; Bxn; *K. ozaenae.* An asterisk indicates the active cysteine residue

4 Carbonyl Reductases and Aldehyde Reductases

There is a group of reductases that catalyze the reduction of various aldehydes and ketones [95, 96]. They occur widely in living organisms, and are thought to be involved in the metabolism of biogenic and xenobiotic carbonyl compounds. Among them, aldehyde reductase (EC 1.1.1.2) and aldose reductase (EC 1.1.1.21) have been characterized in some detail, and it has been suggested that they are implicated in the metabolism of aldehydes derived from biogenic amines, in the reduction of sugar aldehydes and in the biosynthesis of ascorbic acid. The occurrence of other members of this reductase family, the so-called carbonyl reductases, has also been found in a variety of tissues. These enzymes apparently mainly differ from aldehyde reductase and aldose reductase in their ability to reduce aromatic ketones and their sensitivity to specific reductase inhibitors. Some of them were purified from mammalian tissues and characterized [97–99]. They are suggested to be the enzymes responsible for the metabolism of xenobiotic carbonyl compounds. For the practical application of these reductases, the stereospecific reduction of carbonyl compounds with enzymes possessing stereospecificities for the production of various optically active alcohols have been investigated [100].

4.1 Screening for the Ability of Stereospecific Reduction of Carbonyl Compounds

Various micro-organisms have been assayed with regard to their reducing ability using several prochiral carbonyl compounds for the synthesis of D-pantothenic acid and so on (Figs. 8 and 9). Ketopantolactone can be converted to pantolactone by a variety of micro-organisms at pH4 to 6 (Fig. 10a). However, the ratios of D-and L-enantiomers of the pantolactone formed are randomly distributed among the strains and almost completely showed no relation to the genera or sources. For example, *Mucor racemosus* produced almost specifically the L-enantiomer. On the other hand, *Mucor javanicus* yielded a racemic mixture. Among 9 strains of *Rhodotorula glutinis*, 5 gave racemic mixtures, 2 gave the L-enantiomer predominantly and the remaining 2 gave the D-enantiomer [101]. When the same screening is performed at pH 7 to 8, under which the substrate ketopantolactone undergoes rapid and spontaneous hydrolysis to ketopantoic acid, a quite different distribution profile of the reducing activity is observed. Most of the micro-organisms which showed high reducing activity produced only the D-enantiomer (Fig. 10b). Many bacteria belonging to the genus *Agrobacterium and Pseudomonas* have been shown to be potential catalysts for this conversion [102]. The reducing ability toward ethyl 2'-ketopantothenate is also widely distributed in micro-organisms, especially in yeasts (Fig. 10c). Several yeast strains belonging to the genus *Candida, Hansenula* and

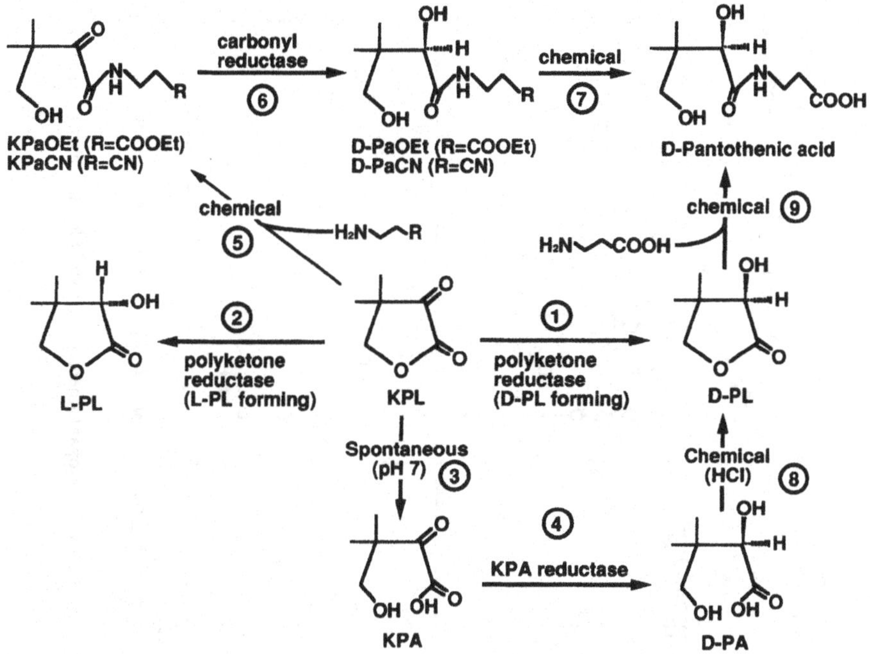

Fig. 8. Possible routes for the synthesis of D-pantothenic acid through the enzymatic transformation. PL, pantolactone; KPL, ketopantolactone; KPA, ketopantoic acid; D-PA, D-pantoic acid; KPaOEt, ethyl 2′-ketopantothenate; KPaCN, 2′-ketopantothenonitrile; D-PaOEt, ethyl D-pantothenate; D-PaCN, D-pantothenonitrile

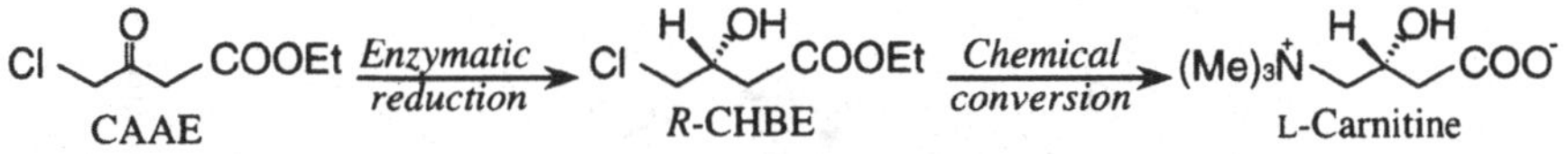

Fig. 9. Chemicoenzymatic route for the synthesis of L-carnitine. CAAE, ethyl 4-chloroacetoacetate; CHBE, ethyl 4-chloro-3-hydroxybutanoate

Sporobolomyces predominantly gave ethyl D-pantothenate, while those belonging to the genus *Saccharomyces*, *Pichia* and *Rhodotorula* gave the L-enantiomer [103]. 2′-Ketopantothenonitrile is also converted to the corresponding alcohol pantothenonitrile; strains belonging to the genera *Sporobolomyces*, *Sporidiobolus* and *Mortierella* specifically gave the D-enantiomer, while *Rhodotorula glutinis* and *Paecilomyces variotii* predominantly produced the L-enantiomer [104].

Reduction of 4-chloroacetoacetate methyl ester usually proceeds stereospecifically. This activity is widely distributed in yeasts, molds and bacteria,

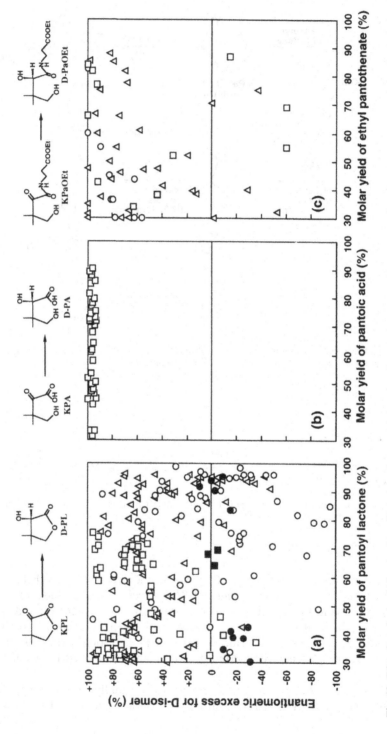

Fig. 10. Diversity of microbial reduction of ketopantolactone (**a**), ketopantoic acid (**b**), and ethyl 2'-ketopantothenate (**c**). Symbols: △, yeasts; ○, molds; □, bacteria; ■, actinomycetes; ●, basidiomycetes

most of which give (S)-enantiomer. *Sporobolomyces salmonicolor*, *Micrococcus luteus* and *Cellulomonas* sp., have been found to produce the (R)-enantiomer predominantly (~ 30% *e.e.*) in high molar conversions [105]. The enzyme from *S. salmonicolor* was characterized and shown to be useful for the practical preparation of the (R)-enantiomer (see Sect. 4.4).

4.2 Novel Carbonyl Reductases and Aldehyde Reductases

The enzyme catalyzing the reduction of ketopantolactone to D-pantolactone was isolated in a crystalline form from the cells of *Candida parapsilosis* and characterized in some detail [106] (see Tables 4 and 5). It is a novel NADPH-dependent carbonyl reductase with a molecular mass of about 40,000. In addition to the reduction of ketopantolactone, the enzyme catalyzes those of a variety of cyclic diketones, including derivatives of ketopantolactone, isatin, camphorquinone and so on, to give the corresponding (R)-alcohols [106, 107]. The enzyme was termed "conjugated polyketone reductase", since the enzyme catalyzes only the reduction of conjugated polyketones as follows.

$$
\begin{array}{cc}
\underset{\substack{\|\\-C-C-}}{O}\ \underset{\substack{\|}}{O} & \longrightarrow \quad \underset{\substack{\|\\-C-C-}}{O}\ \underset{\substack{|}}{H}\ \underset{\substack{|}}{OH}
\end{array}
$$

$$
\underset{\substack{\|\\-C-C=C-C-}}{O}\ \underset{\substack{|}}{X}\ \underset{\substack{|}}{Y}\ \underset{\substack{\|}}{O} \longrightarrow \underset{\substack{|\\-C=C-C=C-}}{OH}\ \underset{\substack{|}}{X}\ \underset{\substack{|}}{Y}\ \underset{\substack{|}}{OH} \qquad \text{[Eq. 4]}
$$

X, Y = H or alkyl

The enzyme yielding the antipode, L-pantolactone, was also isolated from *Mucor ambiguus* [108]. It is also a kind of conjugated polyketone reductase and consists of two polypeptide chains with an identical molecular mass of about 27,500 (see Tables 4 and 5). The occurrence of two kinds of enzymes that show similar substrate specificity but differ from each other in their stereospecificity may be one of the possible reasons why the reduction of ketopantolactone resulted in the formation of the D- and L-enantiomers in varying ratios as shown in Fig. 10a.

The enzyme that catalyzes the stereospecific reduction of ketopantoic acid to D-pantoic acid was isolated in a crystalline form from *Pseudomonas maltophilia* [109] (see Tables 4 and 5). It is an NADPH-dependent enzyme and is strictly specific to ketopantoic acid. The observation that mutants lacking this enzyme require either D-pantoic acid or pantothenic acid for growth and that the revertants regain this activity indicates that the enzyme is ketopantoic acid reductase (EC 1.1.1.169) and is involved in the pantothenate biosynthesis.

The enzyme catalyzing the asymmetric reduction of ethyl 2'-ketopantothenate was isolated from *Candida macedoniensis* [110] (see Tables 4 and 5). The enzyme is NADPH-dependent and shows broad substrate specificity; not only conjugated polyketones, but also aromatic aldehydes and 4-haloacetoacetate esters are reduced. Thus, it has been suggested that it could be a kind of

Table 4. Properties of the carbonyl reductases and aldehyde reductase purified from various microorganisms

	Polyketone reductase (C. parapsilosis)	Polyketone reductase (M. ambiguus)	Ketopantoate reductase (Ps. maltophilia)	Carbonyl reductase (C. macedoniensis)	Aldehyde reductase (S. salmonicolor)
Native M_r	37,000	56,000	116,000	45,000	37,000
Subunit M_r	41,600	27,000	30,500	42,000	37,000
$s_{20,w}$(S)	4.8		7.75		2.85
pI	6.3	6.4	3.5	5.5	4.7
Absorption maxima (nm)	278	276	276	278	278
E(1%, 280 nm)	8.3		20.0		12.2
NH_2 amino acid			Thr	Thr	Val
COOH amino acid			Phe	Thr	Lys
K_m (mM)	0.33 (KPL)	0.71 (KPL)	0.40 (KPA)	2.50 (KPaOEt)	0.36 (CAAE)
V_{max} (mmol/min/mg)	481 (KPL)	541 (KPL)	1,310 (KPA)	120 (KPaOEt)	144 (CAAE)
Cofactor	NADPH	NADPH	NADPH	NADPH	NADPH
Optimum pH	7.0	6.0	6.0	6.5	7.0
Optimum temp. (°C)	40	40	37	40	60
pH stability	6.0–7.5	5.5–7.0	6.0–10	4.5–10.5	6.5–8.5
Thermal stability	42% (40 °C,10 min)	75% (45 °C,10 min)	90% (60 °C,10 min)	100% (55 °C,10 min)	100% (45 °C,10 min)
Inhibitor	quercetin	quercetin			dicoumarol
Reaction mechanism			ordered Bi-Bi		ordered Bi-Bi
Enzyme formation	constitutive	constitutive	constitutive	constitutive	constitutive

See Figs. 8 and 9 for abbreviations

Table 5. Comparison of substrate specificities and stereospecificities of the carbonyl reductases from various micro-organisms

	$-\overset{O}{\underset{\parallel}{C}}-\overset{O}{\underset{\parallel}{C}}-$	R–CHO	X–CHO (o- m- p-)	Ha$\sim\!\!\overset{O}{\parallel}\!\!\sim$COOR	HO$\sim\!\!\overset{O}{\parallel}\!\!\sim$COOH
Polyketone reductase (*C. parapsilosis*)	○ R				
(*M. ambiguus*)	○ S				
Carbonyl reductase (*C. macedoniensis*)	○ R/S=4/1 (KPL)	○		○ S	
Aldehyde reductase (*S. salmonicolor*)		○	○ ○ ○	○ R (CAAE)	
Ketopantoate reductase (*Ps. maltophilia*)					○ R (KPA)

Open circles indicate that enzymes can reduce the substrates indicated in the table. *R* and *S* show the stereospecificities of the enzymes using ketopantolactone (KPL), ethyl 4-chloroacetoacetate (CAAE) or ketopantoate (KPA) as the substrates. Ha = Cl, Br, F or N_3.

carbonyl reductase. The enzyme gives only the D-enantiomer on reduction of ethyl 2′-ketopantothenate, whereas it gives a mixture of the D- and L-enantiomer in a ratio of 4:1 on reduction of ketopantolactone. The product from 4-chloroacetoacetate ethyl ester is (*S*)-4-chloro-3-hydroxybutanoate ethyl ester (95% *e.e.*).

The enzyme giving (*R*)-4-chloro-3-hydroxybutanoate esters was isolated in crystalline form from *S. salmonicolor* and was characterized in some detail [111–113]. It is a kind of aldehyde reductase with a molecular mass of about 37,000. The enzyme absolutely requires NADPH as a cofactor. Besides 4-haloacetoacetate esters, *p*-nitrobenzaldehyde and a variety of other aromatic and aliphatic aldehydes have been found to be reduced well by the enzyme (see Tables 4 and 5). The structural gene of the enzyme was cloned and the enzyme has shown to have similarity to the mammalian aldo-keto-reductase super-family enzymes in primary protein structure [114]. In *S. salmonicolor*, the enzyme comprised 2–6% of the total extractable proteins. Such a high content of the enzyme suggests that this may be one of the reasons why the yeast catalyzes conversion to the (*R*)-enantiomer predominantly.

4.3 Application of Carbonyl Reductases to the Production of Chiral Intermediates for D-Pantothenate Synthesis

At present, commercial production of D-pantothenate depends exclusively on chemical synthesis. The conventional chemical process involves optical resolution of the chemically synthesized racemic pantolactone. To skip the troublesome resolution step, several microbial or enzymatic methods have been proposed. An efficient one-pot synthesis method for ketopantolactone from isobutyraldehyde, sodium methoxide, diethyl oxalate and formalin has been developed [115]. Ketopantolactone is a promising starting material for the synthesis of D-pantolactone because it may permit several microbiological approaches leading to D-pantolactone or D-pantothenate, as shown in Fig. 8. The following reactions have been evaluated for the practical point of view. In every case, washed cells as the catalyst and glucose (or sucrose) as an energy source for the reduction were used. The general principle of the reaction is shown in Fig. 11a.

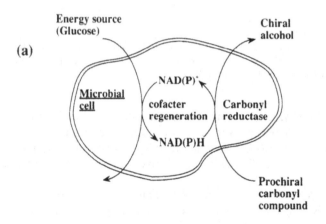

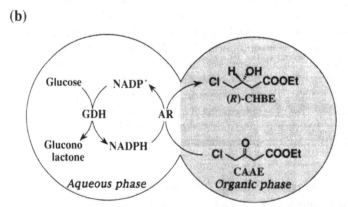

Fig. 11. Priniciple of stereospecific reduction of carbonyl compounds coupled with cofacter regeneration (**a**) and outline of the stereospecific reduction of ethyl 4-chloroacetoacetate (CAAE) by *Sporobolomyces* aldehyde reductase (AR) with glucose dehydrogenase (GDH) as a cofactor regenerator in a water-organic solvent two-phasic system (**b**). CHBE, ethyl 4-chloro-3-hydroxybutanoate

Reduction of ketopantolactone to D-pantolactone (① in Fig. 8). This reaction is catalyzed by conjugated polyketone reductase. About 50 or 90 g/l of D-panto-lactone (98 or 94% *e.e.*, respectively) was produced with a molar yield of nearly 100% on incubation with washed cells of *R. minuta* or *C. parapsilosis*, respectively [115, 116].

Reduction of ketopantoic acid to D-pantoic acid (③, ④ in Fig. 8). *Agrobacterium* sp. S-246 is a good source of ketopantoic acid reductase. The yield of D-pantoic acid reached 119 g/l (molar yield, 90%; optical purity, 98% *e.e.*) on incubation with washed cells of the bacterium [102]. From a practical point of view, ketopantoic acid reduction with *Agrobacterium* cells has several advantages over ketopantoyl lactone reduction with *Candida* (or *Rhodotorula*) cells. The former results in a higher product yield, molar conversion and optical purity of the product than the latter. It is necessary to maintain the substrate level at lower than 3% in the case of the ketopantolactone reduction, but not for the ketopantoic acid reduction.

Reduction of ethyl 2'-ketopantothenate to ethyl 2'-D-pantothenate (⑥ in Fig. 8). The rate of condensation of ketopantolactone or D-pantolactone with ethyl β-alanine, yielding ethyl 2'-ketopantothenate (⑤ in Fig. 8) or ethyl D-panto-thenate, respectively, is quite fast compared to the condensation of ketopanto-lactone or D-pantolactone with β-alanine, and the reaction with ethyl β-alanine proceeds more stoichiometrically [117]. Since the enzymatic hydrolysis of ethyl D-pantothenate has been established [118], if the stereoselective reduction of ethyl 2'-ketopantothenate to ethyl D-pantothenate is possible, both the trouble-some resolution and the incomplete condensation might be avoided at the same time. Carbonyl reductase of *C. macedoniensis* is used for this purpose. Washed cells of the yeast converted ethyl 2'-ketopantothenate (80 g/l) almost specifically to ethyl D-pantothenate (> 98% *e.e.*), with a molar yield of 97.2% [103]. In a similar manner, 2'-ketopantothenonitrile (50 g/l) was converted to D-pan-tothenonitrile (93.6% *e.e.*), with a molar yield of 95.6%, on incubation with *Sporidiobolus salmonicolor* cells as a catalyst [104].

4.4 Application of Aldehyde Reductase to the Production of Chiral Building Blocks for L-Carnitine Synthesis

Aldehyde reductase of *S. salmonicolor* can catalyze the asymmetric reduction of 4-haloacetoacetate esters to the corresponding (*R*)-alcohols, which are promis-ing chiral building blocks for the preparation of a variety of optically active compounds such as L-carnitine [119] (Fig. 9).

Practical preparation is carried out in a water–organic solvent two-phase system as shown in Fig. 11b, because both substrate and product strongly inhibit enzyme reactions. Crude extracts from *S. salmonicolor* after heat- and acetone-treatments and glucose dehydrogenase of *Bacillus megaterium* are used as a source of aldehyde reductase and a cofactor regenerator. *n*-Butylacetate is the most suitable organic solvent. It shows high partition efficiencies toward both the substrate and product, and both of the enzymes are stable in the presence of this organic solvent. In a 1.6 l–1.6 l bench-scale two-phase reaction, 134 g ethyl (*R*)-4-chloro-3-hydroxybutanoate (86% *e.e.*) was produced from the corresponding keto ester in a molar yield of 95.4% [105, 120].

Recently, this two-phase reaction was improved by using an *Esherichia coli* transformant expressing both genes for aldehyde reductase and glucose dehydrogenase as the catalyst. When the *E. coli* transformant cells were incubated in a two-phase system, 300 mg/ml of the substrate was almost stoichiometrically converted to ethyl (*R*)-4-chloro-3-hydroxybutanoate (92% *e.e.*) in 16 h [114].

5 Lactonases

Lactonases reversibly catalyze the hydrolysis of lactone compounds into hydroxy acids, that is, they mediate the interconversion between the lactone and acid forms of hydroxy carboxylic acids. Gluconolactonase (aldonolactonase; EC 3.1.1.17), L-arabinonolactonase (EC 3.1.1.15) and D-arabinonolactonase (EC 3.1.1.30) specifically hydrolyze aldonate lactones to the respective aldonic acids. The activities of these enzymes have been detected in various organisms [121–128]. As to the roles of these enzymes in vivo, it has been suggested that gluconolactonase participates in the formation of L-gulono-γ-lactone from L-gulonate in L-ascorbate biosynthesis (Fig. 12) [123]. L-Rhamnonolactonase (EC 3.1.1.65) and D-xylonolactonase (EC 3.1.1.68) of *Pullularia pullulans* and *Pseudomonas fragi*, respectively, are involved in the oxidative degradation of L-rhamnose and D-xylose, respectively (Fig. 12) [129, 130]. In addition to these enzymes, dihydrocoumarin hydrolase (EC 3.1.1.35) catalyzing the hydrolysis of aromatic lactones [131], γ-lactonase (EC 3.1.1.25) catalyzing the hydrolysis of aliphatic γ-lactones [132, 133] and so on have been reported.

5.1 Novel Microbial Lactonases

During the course of studies on the microbial production of chiral intermediates for D-pantothenic acid [2, 134, 135], Shimizu and co-workers found that several micro-organisms, such as *Fusarium*, *Brevibacterium* and so on, produce a novel enzyme that catalyzes the hydrolysis of aldonate lactones or aromatic lactones [136, 137].

The enzyme isolated from the cells of *Fusarium oxysporum* [136] has the relative molecular mass of 125 kDa and the subunit molecular mass of 60 kDa. The enzyme thus appears to be a dimer consisting of identical subunits. About

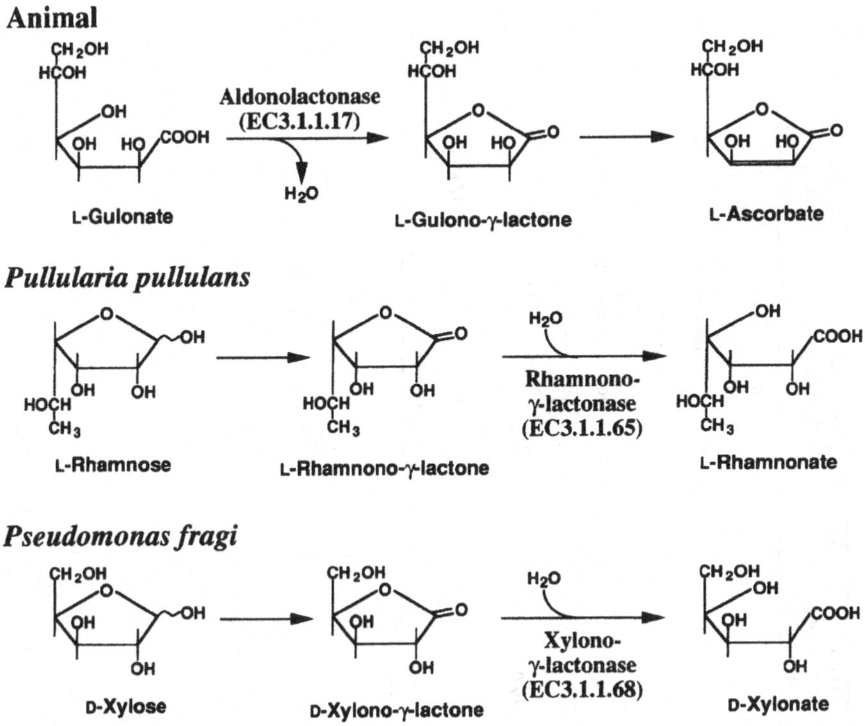

Animal

L-Gulonate → (Aldonolactonase (EC3.1.1.17), H_2O) → L-Gulono-γ-lactone → L-Ascorbate

Pullularia pullulans

L-Rhamnose → L-Rhamnono-γ-lactone → (H_2O, Rhamnono-γ-lactonase (EC3.1.1.65)) → L-Rhamnonate

Pseudomonas fragi

D-Xylose → D-Xylono-γ-lactone → (H_2O, Xylono-γ-lactonase (EC3.1.1.68)) → D-Xylonate

Fig. 12. Lactonase reactions involved in ascorbate biosynthesis and degradation of L-rhamnose and D-xylose

1 mol of calcium per subunit and 15.4% (w/w) glucose equivalent of carbohydrate are included in the enzyme. Calcium is necessary for both enzyme activity and stability. The enzyme carries three kinds of N-linked high-mannose-type sugar chains at the 28th, 106th 179th and 277th Asn residues of the enzyme polypeptide. The carbohydrate moiety is an essential participant in the stabilization of the enzyme. The enzyme hydrolyzes aldonate lactones, such as D-galactono-γ-lactone, L-mannono-γ-lactone, D-gulono-γ-lactone and D-glucono-δ-lactone, stereospecifically, but the corresponding enantiomers (i.e., L-galactono-γ-lactone, D-mannono-γ-lactone, L-gulono-γ-lactone and L-glucono-δ-lactone, respectively) do not serve as substrates. According to Haworth's projection system, all of these substrates have downward hydroxy groups at their 2-position carbons in common (Table 6), while their enantiomers have upward hydroxy groups at their 2-position carbons. The enzyme can also catalyze the asymmetric hydrolysis of D-pantolactone. For every substrate, the reverse reaction, i.e., the lactonization of aldonic acids, is catalyzed under acidic pH conditions. Aromatic lactones, such as dihydrocoumarin, homogentisic acid lactone, 2-coumaranone and 3-isochromanone, are also effective substrates, but lactonization does not take place with these aromatic substrates. They are good

Table 6. Substrate specificity of lactonase from *F. oxysporum*

Substrate (150 mM)	Relative hydrolyzing activity (%)	Km (mM)	Substrate (150 mM)	Relative hydrolyzing activity (%)	Km (mM)	Substrate (2.5 mM)	Relative hydrolyzing activity (%)	Km (mM)
D-Galactono-γ-lactone	100	3.6	α,β-Glucooctanoic γ-lactone	16.4	—	Dihydrocoumarin	100	6.3
L-Mannono-γ-lactone	128	23	D-Ribono-γ-lactone	39.0	2.5	Homogentisic acid lactone	16.6	2.5
D-Gulono-γ-lactone	167	29	D-Erythrono-γ-lactone	117	377	2-Coumaranone	16.4	8.7
D-glycero-D-gulo-Heptono-γ-lactone	59.4	118	D-Glucono-δ-lactone	107	—	3-Isochromanone	0.58	4.4
D-glycero-L-manno-Heptono-γ-lactone	132	167	D-Pantolactone	48.3	120			

substrates for dihydrocoumarin hydrolase and have no hydroxy group at their 2-position carbons [131]. With regard to the reverse reactions, no detectable lactone compounds were formed.

Brevibacterium protophormiae produces a similar kind of lactonase, which hydrolyzes various kinds of aromatic lactones as well as the *Fusaruim* lactonase, but aldonate lactones are not hydrolyzed. The enzyme hydrolyzes only L-pantolactone; D-pantolactone is not a substrate. The relative molecular mass of the native enzyme is 62 kDa, and the subunit molecular mass is 26.5 kDa [137].

5.2 Optical Resolution of Racemic Pantolactone with Lactonase

The principle of the optical resolution of racemic pantolactone is shown in Fig. 13. If racemic pantolactone is used as a substrate for the hydrolysis reaction by the stereospecific lactonase, only the D- or L-pantolactone might be converted to D- or L-pantoic acid and the L- or D-enantiomer might remain intact, respectively. Consequently, the racemic mixture could be resolved into D-pantoic acid and L-pantolactone, or D-pantolactone and L-pantoic acid. In the case of L-pantolactone-specific lactonase, the optical purity of the remaining D-pantolactone might be low, except when the hydrolysis of L-pantolactone is complete. On the other hand, using the D-pantolactone-specific lactonase, D-pantoic acid with high optical purity could be constantly obtained independently of the hydrolysis yield. Therefore, the enzymatic resolution of racemic pantolactone with D-pantolactone-specific lactonase was investigated [138–140].

Distribution of D-pantolactone-specific lactonase is narrow; only filamentous fungi of three genera, i.e., *Fusarium*, *Gibberella* and *Cylindrocarpon*, show high hydrolytic activity. Washed cells of these fungi, as well as the purified lactonase, catalyze well this stereospecific hydrolysis. For example, D-pantolactone in a racemic mixture of pantolactone (700 mg/ml) was almost

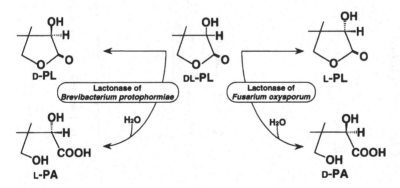

Fig. 13. Novel microbial lactonases catalyzing stereospecific hydrolysis of pantolactone. PL, pantolactone; PA, pantoic acid

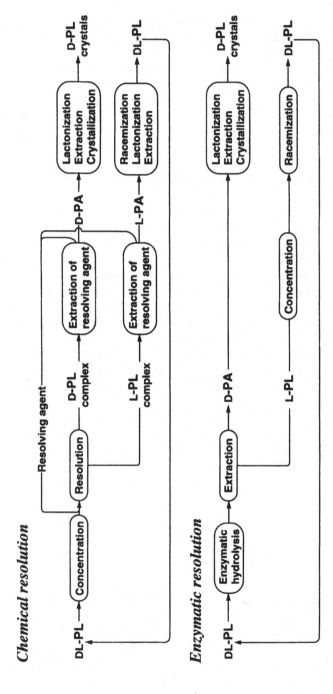

Fig. 14. Comparison of enzymatic and conventional chemical resolution processes for DL-pantolactone. PL, pantolactone; PA, pantoic acid

stoichiometrically hydrolyzed to D-pantoic acid on incubation with *F. oxysporum* cells for 24 h under conditions in which the pH was automatically maintained at 7.0 (optical purity for D-pantoic acid, 96% *e.e.*). In practice the hydrolysis of the D-pantolactone in a racemic mixture is achieved by immobilized cells of *F. oxysporum* as the catalyst. Stable catalysts with high hydrolytic activity can be prepared by entrapping the fungal cells into calcium alginate gels. When the immobilized cells were incubated in a reaction mixture containing 350 g/l DL-pantolactone for 21 h at 30 °C under the conditions of automatic pH control (pH 6.8–7.2), 90–95% of the D-pantolactone was hydrolyzed (optical purity, 90–97% *e.e.*). After repeated reactions for 180 times (i.e., 180 days), the immobilized mycelia retained more than 90% of their initial activity.

The overall process for this enzymatic resolution is compared with the conventional chemical process in Fig. 14. The enzymatic process can skip several tedious steps which are necessary in chemical resolution and this is a considerable practical advantage. There have been several reports on the application of enzymatic asymmetric hydrolysis to the optical resolution of pantolactone [141, 142]. In these cases, esterified substrates, such as *O*-acetyl or *O*-formyl pantolactone, and lipases were used as the starting materials and catalysts, respectively. Since the lactonase of *F. oxysporum* hydrolyzes the intramolecular ester bond of pantolactone, it is not necessary to modify the substrate, pantolactone. This is one of the practical advantages of this enzyme.

6 Fatty Acid Transforming Enzyme Systems

C_{20} Polyunsaturated fatty acids (PUFAs) such as Mead acid (20:3ω9), dihomo-γ-linolenic acid (DHGA, 20:3ω6), arachidonic acid (AA, 20:4ω6), and 5, 8, 11, 14, 17-*cis*-eicosapentaenoic acid (EPA, 20:5ω3) (see Fig. 15, for structures and biosynthetic pathways) exhibit unique biological activities, such as lowering of the plasma cholesterol level, prevention of thrombosis and so on. Also, they are precursors for large family of prostaglandin-related C_{20} compounds [143–145]. The inclusion of EPA and 4, 7, 10, 13, 16, 19-*cis*-docosahexaenoic acid (DHA) as dietary supplements has been recommended for the prevention of heart diseases [146], and γ-linolenic acid (18:3ω6) has been recommended for the relief of eczema [147]. Accordingly, PUFAs are highly important substances in the pharmaceutical, medical and nutritional fields. Since food sources rich in these PUFA are limited to a few seed oils which contain 18:3ω6 and fish oils which contains EPA and DHA, recent investigations have been focused on micro-organisms as alternative sources of these PUFAs. Recently, several fungi were found to produce large amount of lipids rich in DHGA, AA or EPA, or all of these C_{20} PUFAs. They are new and promising sources of C_{20} PUFAs.

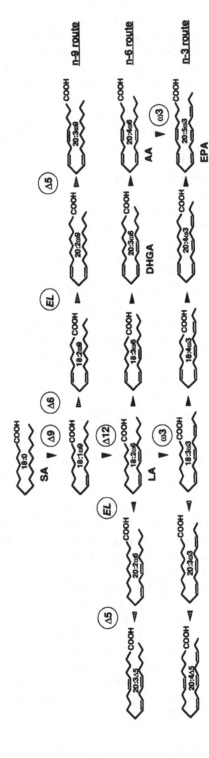

Fig. 15. Proposed biosynthetic pathways for PUFA biosynthesis in an arachidonic acid-producing fungus, *M. alpina* 1S-4: Broken arrows show by-paths through which ω9 fatty acids and nonmethylene-interrupted PUFAs are formed by mutants of *M. alpina* 1S-4, Mut48 and Mut49, respectively. Δn, Δn desaturation; EL, elongation

6.1 Arachidonic Acid-Producing Micro-organisms

Most C_{20} PUFA producers were found to be filamentous fungi belonging to the orders Mucorales and Entomophthorales. Mainly they produce C_{20} PUFAs of the ω6 family (i.e., AA and DHGA) together with C_{18} PUFA of the same family (i.e., 18:3ω6). In particular, several strains of the genus *Mortierella* have been shown to be potential producers of these PUFAs. All of the *Mortierella* strains found as C_{20} PUFA producers belong to the subgenus *Mortierella*. Neither the stock cultures nor isolates belonging to the subgenus *Micromucor* showed any detectable accumulation of C_{20} PUFA, although they were good producers for 18:3ω6 [148]. As a result of this screening effort, a soil-isolated fungus, which was taxonomically identified as *Mortierella alpina* 1S-4, was obtained as a potent producer of AA [149, 150]. The fungus produces 30–50 kg/kl of mycelia (dry wt) containing about 60% lipids (mainly triglyceride), and 40–60% of AA in the total fatty acids, under large-scale fermentation involving intermittent feeding of glucose and 7 days cultivation at 28 °C (Table 7). Thus, the resultant fungal mycelia are a rich source of AA.

6.2 Enzymes Involved in Fatty Acid Transformation

Figure 15 shows the biosynthetic pathway of ω9, ω6 and ω3 PUFAs from stearic acid (18:0). In *M. alpina* 1S-4, AA is synthesized through the n-6 route where Δ9, Δ12, Δ6 and Δ5 desaturases and elongase function. 70–90% of synthesized AA is accumulated as triglyceride in fungal mycelia. In *M. alpina* 1S-4, Δ9, Δ12, Δ6 and Δ5 desaturases and elongase, which are involved in the biosynthesis of ω6 PUFAs through the n-6 route, can also act on the corresponding PUFAs in n-9 and n-3 routes. Consequently, *M. alpina* 1S-4 can produce ω9 and ω3 PUFAs when incubated with suitable intermediates of these routes. Actually, EPA was produced through the n-3 route from 18:3ω3 when the fungus was grown in media containing linseed oil in which 18:3ω3 amounts to about 60% of the total fatty acids [151] (see also Table 7). The wide substrate specificity of desaturases and elongase were also confirmed through synthetic studies of odd numbered PUFAs (i.e, 19:3ω5 and 19:4ω6), and those with an ω-terminal double bond (i.e., 20:3ω1 and 20:4ω1) [152, 153].

The ability of *M. alpina* 1S-4 to convert added precursor fatty acids, such as 18:3ω3, to corresponding C_{20} PUFAs is very promising from a viewpoint of biotransformation of various kinds of naturally occurring or chemically synthesized C_{18} fatty acids.

6.3 Mutants Defective in Fatty Acid Desaturation

To extend this unique fungal ability of fatty acid transformation, several kinds of PUFA biosynthesis mutants of *M. alpina* 1S-4 have been isolated through

Table 7. PUFA production by *M. alpina* 1S-4 and its mutants

PUFA	Major source in food	Biological function, cellular distribution, use, etc.	Producer	Lipid content		PUFA productivity		Other remarks
				Total FA (mg/g dry mycelia)	%PUFA content in total FA	PUFA content (mg/g dry mycelia)	PUFA yield (g/l)	
ω6 PUFA Dihomo-γ-linolenic acid (20:3ω6, DHGA)	not known [mother's milk (0.3%)]	Precursor of PG 1	Wild strain	463	23.1	107	2.17	Inhibition of Δ5 DS[a] by sesamin
			Mutant Mut44	488	42	205	4.1	Accumulation by Δ5 DS mutant
Arachidonic acid (20:4ω6)	Animal meat (2–3%)	Major component of membrane phospholipids, precursor of PG2	Wild strain	400–500	30–70	100–200	3–5	High AA content oil
ω9 PUFA Eicosadienoic acid (20:2ω9)	not known	not known	Mutant 226-9	304	22	62	0.7	Conversion of OA[b] through n-9 route by Δ12 DS and Δ5 DS double mutant
Mead acid (20:3ω9, MA)	not known	Precursor of LT3	Mutant Mut48	427	33	141	1.9	Conversion of OA through n-9 route by Δ12 DS mutant
ω3 PUFA Eicosatetraenoic acid (20:4ω3)	not known	Precursor of 17, 18-dehydro-PGE1	Mutant S14	574	12	67	1.7	Conversion of ALA[c] through n-3 route by Δ5 DS mutant
Eicosapentaenoic acid (20:5ω3, EPA)	Fish, shellfish algae, sardine oil (10%), cod liver oil (10%)	Precursor of PG3, pharmaceuticals, food additive	Wild strain	585	7.1	42	1.35	Conversion of ALA through n-3 route at normal growth temperature
			Mutant Mut48	320	20	64	1.0	Normal growth temperature by Δ12 DS mutant Low w6 PUFA

[a] DS: desaturase.
[b] OA: 18:1ω9, oleic acid
[c] ALA: 18:3ω3, α-linolenic acid

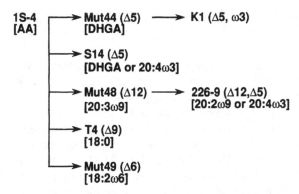

Fig. 16. Mutants derived from an arachidonic acid-producing fungus, *M. alpina* 1S-4: Desaturases missing are shown in parentheses. Major fatty acids produced are shown in brackets

mutagenesis of the parental spores [154, 155] (Fig. 16). The properties of some mutants are as follows.

Δ5 Desaturase-defective mutants, Mut44 and S14, are characterized by a high DHGA level and a reduced AA level [156, 157]. Thus, these mutants are potent producers of DHGA.

Mutant Mut49 was considered to be defective in Δ6 desaturase because of a high 18:2ω6 level and low levels of 18:3ω6, DHGA and AA in its mycelia. In addition to the accumulation of 18:2ω6, two nonmethylene-interrupted PUFAs, i.e., 5, 11, 14-*cis*-eicosatrienoic acid (20:3Δ5) and 5, 11, 14, 17-*cis*-eicosatetraenoic acid (20:4Δ5), were found [158]. These PUFAs are considered to be converted from 18:2ω6 or 18:3ω3, respectively, by three subsequent reactions, i.e., Δ6 desaturation, elongation and Δ5 desaturation (Fig. 15).

Mutant T4 is considered to be defective in Δ9 desaturation [155]. Its mycelial fatty acids included 38% stearic acid (18:0) mainly as free fatty acids, the level being only 5% for the wild type. Since the fatty acid composition of phosphatidylethanolamine and phosphatidylcholine at *sn*-2 position included more than 75% of PUFA, the accumulation of free fatty acids was supposed to be due to the lack of PUFAs.

Mutant Mut48 is characterized by a high oleic acid (18:1ω9) level and the absence of ω6 PUFAs in its mycelia. In contrast to the wild type, this mutant produces three ω9 PUFAs, i.e., 6, 9-*cis*-octadecadienoic acid (18:2ω9), 8, 11-*cis*-eicosadienoic acid (20:2ω9), and 20:3ω9 [159]. The formation of 20:3ω9 would occur through the desaturation of 18:1ω9 at Δ6 position into 18:2ω9 followed by elongation and Δ5 desaturation to 20:3ω9 (Fig. 15).

Mutant 226-9, which was derived from strain Mut48, is defective in both Δ12 and Δ5 desaturases. The mutant produces 20:2ω9 and 20:4ω3 from 18:1ω9 and 18:3ω3, respectively [160].

6.4 Fatty Acid Transformation by Enzyme Systems of M. alpina 1S-4

Using the parental and mutant strains of M. alpina 1S-4 as fatty acid transforming enzyme systems, ω6, ω3 or ω9 PUFAs are produced from the corresponding precursors, i.e., 18:2ω6, 18:3ω3 or 18:1ω9, respectively (Table 7).

From 18:2ω6, parental strain or Mut44 (Δ5 desaturase mutant) produces a triglyceride with high AA content or one with high DHGA content, respectively, On cultivation for 6 days at 28 °C, Mut44 produced 3.2 g DHGA per l of culture broth (123 mg/g dry mycelia), which accounted for 23.4% of the total mycelial fatty acids [156]. The mycelial AA amounted to only 19 mg/g dry mycelia (0.5 g/l culture broth) because of defective Δ5 desaturase, which accounted for 3.7% of the total mycelial fatty acids.

From 18:1ω9, Mut48 (Δ12 desaturase mutant) produces 20:3ω9 [161]. On cultivation at 20 °C for 10 days, the production of 20:3ω9 by Mut48 reached 0.8 g/l (56 mg/g dry mycelia), accounting for 15% of the total mycelial fatty acids. Similarly a mutant 226-9 (Δ12 and Δ5 desaturase double mutant) produces 1.9 g 20:2ω9/l of culture broth (101 mg/g dry mycelia; 20.0% of the total fatty acids) [161].

From 18:3ω3, parental strain or Mut48 (Δ12 desaturase mutant) produces EPA efficiently. Especially, Mut48, when incubated with 18:3ω3 (i.e. linseed oil) produces a unique triglyceride rich in ω3 PUFAs with a low amount of ω6 PUFA (AA), because the mutant cannot produce 18:2ω6, the parental fatty acid of the ω6 PUFAs [162]. On cultivation at 20 °C for 10 days in a medium supplemented with linseed oil, the production of EPA by Mut48 was 1 g/l culture broth (64 mg/g dry mycelia), accounting for 20% of the total mycelial fatty acids. The mycelial AA content was 26 mg/g dry mycelia (0.4 g/l), accounting for 7.8% of the total mycelial fatty acids (This was from 18:2ω6 in linseed oil). For the production of 20:4ω3, S14 (Δ5 desaturase mutant) or 226-9 (Δ12 and Δ5 double mutant) has been shown to be useful [163].

7 Conclusion

Compared to chemical catalysts, enzymes have many superior properties. One of these is their substrate specificity. Enzymes catalyze specific reactions involving only one compound or a few that are structurally related, and they distinguish almost completely between stereoisomers and regioisomers. Therefore, only a very specific change in a functional group or bond of a compound is accelerated by a given enzyme. As a result, one single product may be expected as long as only one enzyme is involved in a reaction. This property enables us to synthesize complex and optically active compounds which are biologically and chemically useful, but makes it difficult to find a suitable enzyme for the reaction of interest. The most efficient and successful way to find the required enzymes is to screen them in micro-organisms. As we have presented here about microbial enzymes, the specific features of the microbial world are its unusual diversity and versatility. In comparison to higher cells, micro-organisms show a great ability to adapt to a wide variety of environments. Therefore, they have considerable potential for inducing new or novel enzyme systems capable of converting foreign substrates. It is possible to obtain and cultivate micro-organisms that can survive or grow in extraordinary environments, e.g., psychrophilic, thermophilic, acidophilic and alkalophilic ones. These micro-organisms are capable of producing unique enzymes stable towards heat, alkalis, and acids. In addition, techniques for the cultivation and purification of enzymes, as well as gene and enzyme technology, have made great advances. Taking all of these facts into consideration, there must be far more ways than one can imagine of using microbial enzymes for the production of biologically and chemically useful compounds.

Acknowledgements. This work was carried out in the laboratory of Fermentation Physiology and Applied Microbiology, Department of Agricultural Chemistry, Kyoto University. We greatly appreciate the efforts of all members of the laboratory and its collaborators. In particular, we are deeply indebted to Emeritus Prof. Hideaki Yamada of Kyoto University, a former professor of this laboratory and a pioneer of screening for novel microbial enzymes, for his valuable discussion and encouragement.

8 References

1. Yamada H, Shimizu S (1985) Ullmann's Encyclopedia of Industrial Chemistry, Vol. A4. VCH Verlagsgesellschaft, Weinheim, p 150
2. Yamada H, Shimizu S (1988) Angew Chem Int Ed Engl 27: 622
3. International Union of Biochemistry (1992) Enzyme nomenclature 1993. Academic Press, New York
4. Yamada H, Takahashi S, Kii Y, Kumagai H (1978) J Ferment Technol 56: 484
5. Takahasni S, Kii Y, Kumagai H, Yamada H (1978) J Ferment Technol 56: 492
6. Shimizu S, Shimada H, Takahashi S, Ohashi T, Tani Y, Yamada H (1980) Agric Biol Chem 44: 2233
7. Ogawa J, Shimizu S, Yamada H (1993) Eur J Biochem 212: 685
8. Ogawa J, Chung MCM, Hida S, Yamada H, Shimizu S (1994) J Biotechnol 38: 11
9. Takahashi S (1994) Abstract of the 8th German–Japanese Workshop on Enzyme Technology, Tayama, Japan: 23
10. Ogawa J, Honda M, Soong CL, Shimizu S (1995) Biosci Biotech Biochem 59: 1960
11. Kim JM, Shimizu S, Yamada H (1987) Biochem Biophys Res Commun 142: 1006
12. Shimizu S, Kim JM, Yamada H (1989) Clin Chim Acta 185: 241
13. Yamada H, Shimizu S, Kim JM, Shinmen Y, Sakai T (1985) FEMS Microbiol Lett 30: 337
14. Shimizu S, Kim JM, Shinmen Y, Yamada H (1986) Arch Microbiol 145: 322
15. Kim JM, Shimizu S, Yamada H (1986) J Biol Chem 261: 11832

16. Kim JM, Shimizu S, Yamada H (1987) Arch Microbiol 147: 58
17. Kim JM, Shimizu S, Yamada H (1987) FEBS Lett 210: 77
18. Kim JM, Shimizu S, Yamada H (1986) Agric Biol Chem 50: 2811
19. Kim JM, Shimizu S, Yamada H (1987) Agric Biol Chem 51: 1167
20. Ogawa J, Kim JM, Nirdnoy W, Amano Y, Yamada H, Shimizu S (1995) Eur J Biochem 229: 284
21. Ogawa J, Nirdnoy W, Yamada H, Shimizu S (1995) Biosci Biotech Biochem 59: 1737
22. Ogawa J, Shimizu S (1994) Eur J Biochem 223: 625
23. Matthews MM, Liao W, Kvalnes-Krick KL, Traut TW (1992) Arch Biochem Biophys 293: 254
24. Tamaki N, Mizutani N, Kikugawa M, Fujimoto S, Mizota C (1987) Eur J Biochem 169: 21
25. Wasternack C, Lippmann G, Reinbotte H (1979) Biochim Biophys Acta 570: 341
26. Campbell LL (1960) J Biol Chem 235: 2375
27. Syldatk C, Cotoras D, Dombach G, Groß C, Kallwaß H, Wagner F (1987) Biotechnol Lett 9: 25
28. Sano K, Yokozeki K, Eguchi C, Kagawa T, Noda I, Mitsugi K (1977) Agric Biol Chem 41: 819
29. Ishikawa T, Watabe K, Mukohara Y, Kobayashi S, Nakamura H (1993) Biosci Biotech Biochem 57: 982
30. Yamashiro A, Yokozeki K, Kano H, Kubota K (1988) Agric Biol Chem 52: 2851
31. Ishikawa T, Mukohara Y, Watabe K, Kobayashi S, Nakamura H (1994) Biosci Biotech Biochem 58: 265
32. Ogawa J, Miyake H, Shimizu S (1995) Appl Microbiol Biotechnol 43: 1039
33. Runser SM, Meyer PC (1993) Eur J Biochem 213: 1315
34. Watabe K, Ishikawa T, Mukohara Y, Nakamura H (1992) J Bacteriol 174: 962
35. Ishikawa T, Mukohara Y, Watabe K, Kobayashi S, Nakamura H (1994) Biosci Biotech Biochem 58: 265
36. Syldatk C, Wagner F (1990) Food Biotechnol 4: 87
37. Watabe K, Ishikawa T, Mikohara Y, Nakamura H (1992) J Bacteriol 174: 7989 Biotech Biochem 59: 2292
38. Ogawa J, Nirdnoy W, Tabata M, Yamada H, Shimizu S (1995) Biosci Biotech Biochem 59: 2292
39. Kobayashi M, Nagasawa T, Yamada H (1992) Trends Biotechnol 10: 402
40. Conn EE (1981) in: Vennesland B, Conn EE, Knowles CJ, Westley J, Wissinic F (eds.) Cyanide in Biology. Academic Press, London, p. 183
41. Kobayashi M, Shimizu S (1994) FEMS Microbiol Lett 120: 217
42. Nagasawa T, Yamada H (1989) Trends Biotechnol 7: 153
43. Jallageas J-C, Arnaud A, Galzy P (1980) in: Fiechter A (ed) Advances in Biochemical Engineering. Springer-Verlag, Berlin, p. 1 (vol. 14)
44. Thompson LA, Knowles CJ, Linton EA, Wyatt JM (1988) Chemistry in Britain September: 900
45. Yamada H, Kobayashi M (1996) Biosci Biotech Biochem 60: 1391
46. Asano Y, Fujishiro K, Tani Y, Yamada H (1982) Agric Biol Chem 46: 1165
47. Asano Y, Yasuda T, Tani Y, Yamada H (1982) Agric Biol Chem 46: 1183
48. Yamada H, Ryuno K, Nagasawa T, Enomoto K, Watanabe I (1986) Agric Biol Chem 50: 2859
49. Ryuno K, Nagasawa T, Yamada H (1988) Agric Biol Chem 52: 1813
50. Nagasawa T, Nanba H, Ryuno K, Takeuchi K, Yamada H (1987) Eur J Biochem 162: 691
51. Sugiura Y, Kuwahara J, Nagasawa T, Yamada H (1987) J Am Chem Soc 109: 5848
52. Sugiura Y, Kuwahara J, Nagasawa T, Yamada H (1988) Biochem Biophys Res Commun 154: 522
53. Nagasawa T, Yamada H (1987) Biochem Biophys Res Commun 147: 701
54. Honda J, Nagamune T, Teratani Y, Hirata A, Sasabe H, Endo I (1992) Ann NY Acad Sci 672: 29
55. Nagamune T, Kurata H, Hirata M, Honda J, Hirata A, Endo I (1990) Photochem Photobiol 51: 87
56. Honda J, Kandori H, Okada T, Nagamune T, Shichida Y, Sasabe H, Endo I (1994) Biochemistry 33: 3577
57. Nagamune T, Honda J, Cho W-D, Kamiya N, Teratani Y, Hirata A, Sasabe H, Endo I (1991) J Mol Biol 220: 221
58. Nagasawa T, Takeuchi K, Yamada H (1991) Eur J Biochem 196: 581
59. Komeda H, Kobayashi M, Shimizu S (1996) Proc Natl Acad Sci USA 93: 4267
60. Komeda H, Kobayashi M, Shimizu S (1996) J Biol Chem 271: 15796

61. Kobayashi M, Suzuki T, Fujita T, Masuda M, Shimizu S (1995) Proc Natl Acad Sci USA 92: 714
62. Kobayashi M, Fujita T, Shimizu S (1996) Appl Microbiol Biotechnol 45: 176
63. Kobayashi M, Nishiyama M, Nagasawa T, Horinouchi S, Beppu T, Yamada H (1991) Biochim Biophys Acta 1129: 23
64. Nishiyama M, Horinouchi S, Kobayashi M, Nagasawa T, Yamada H, Beppu T (1991) J Bacteriol 173: 2465
65. Ikehata O, Nishiyama M, Horinouchi S, Beppu T (1989) Eur J Biochem 181: 563
66. Hashimoto Y, Nishiyama M, Horinouchi S, Beppu T (1994) Biosci Biotech Biochem 58: 1859
67. Duran R, Nishiyama M, Horinouchi S, Beppu T (1993) Biosci Biotech Biochem 57: 1323
68. Mayaux J-F, Cerbelaud E, Soubrier F, Faucher D, Petre D (1990) J Bacteriol 172: 6764
69. Mayaux J-F, Cerbelaud E, Soubrier F, Yeh P, Blanche F, Petre D (1991) J Bacteriol 173: 6694
70. Kobayashi M, Komeda H, Nagasawa T, Nishiyama M, Horinouchi S, Beppu T, Yamada H, Shimizu S (1993) Eur J Biochem 217: 327
71. Hashimoto Y, Nishiyama M, Yu F, Watanabe I, Horinouchi S, Beppu T (1992) J Gen Microbiol 138: 1003
72. Nagasawa T, Shimizu H, Yamada H (1993) Appl Microbiol Biotechnol 40: 189
73. Nagasawa T, Mathew CD, Mauger J, Yamada H (1988) Appl Environ Microbiol 54: 1766
74. Thimann KV, Mahadevan S (1964) Arch Biochem Biophys 105: 133
75. Mahadevan S, Thimann KV (1964) Arch Biochem Biophys 107: 62
76. Bartling D, Seedorf M, Mithofer A, Weiler EW (1992) Eur J Biochem 205: 417
77. Bartling D, Seedorf M, Schmidt RC, Weiler EW (1994) Proc Natl Acad Sci USA 91: 6021
78. Bartel B, Fink GR (1994) Proc Natl Acad Sci USA 91: 6649
79. Nagasawa T, Kobayashi M, Yamada H (1988) Arch Microbiol 150: 89
80. Kobayashi M, Nagasawa T, Yamada H (1989) Eur J Biochem 182: 349
81. Kobayashi M, Yanaka N, Nagasawa T, Yamada H (1990) J Bacteriol 172: 4807
82. Kobayashi M, Yanaka N, Nagasawa T, Yamada H (1991) FEMS Microbiol Lett 77: 121
83. Kobayashi M, Komeda H, Yanaka N, Nagasawa T, Yamada H (1992) J Biol Chem 267: 20746
84. Kobayashi M, Yanaka N, Nagasawa T, Yamada H (1992) Biochemistry 31: 9000
85. Kobayashi M, Izui H, Nagasawa T, Yamada H (1993) Proc Natl Acad Sci USA 90: 247
86. Mathew CD, Nagasawa T, Kobayashi M, Yamada H (1988) Appl Environ Microbiol 54: 1030
87. Kobayashi M, Nagasawa T, Yanaka N, Yamada H (1989) Biotechnol Lett 11: 27
88. Kobayashi M, Yanaka N, Nagasawa T, Yamada H (1990) J Antibiotics 43: 1316
89. Kobayashi M, Nagasawa T, Yamada H (1988) Appl Microbiol Biotechnol 29: 231
90. Kobayashi M, Yanaka N, Nagasawa T, Yamada H (1990) Tetrahedron 46: 5587
91. Bengis-Garber C, Gutman AL (1988) Tetrahedron Lett 29: 2589
92. Bengis-Garber C, Gutman AL (1989) Appl Microbiol Biotechnol 32: 11
93. Stalker DM, McBride KE, Malyj LD (1988) Science 242: 419
94. Stalker DM, Malyj LD, McBride KE (1988) J Biol Chem 263: 6310
95. Ernster L (1975) Method Enzymol 10: 309
96. Martius C (1963) in: Boyer PD (ed) The enzymes, 2nd edn, vol 7, Academic Press, New York, p. 517
97. Hojberg BO, Blomberg K, Stenberg S, Lind C (1981) Arch Biochem Biophys 207: 205
98. Lind C, Hojberg BO (1981) Arch Biochem Biophys 207: 217
99. Hosoda S, Nakamura W, Hayashi K (1974) J Biol Chem 249: 6416
100. Hummel W, Kula MR (1989) Eur J Biochem 184: 1
101. Shimizu S, Hata H, Yamada H (1984) Agric Biol Chem 51: 2285
102. Kataoka M, Shimizu S, Yamada H (1990) Agric Biol Chem 54: 177
103. Kataoka M, Shimizu S, Doi Y, Yamada H (1990) Appl Environ Microbiol 56: 3595
104. Kataoka M, Shimizu S, Doi Y, Sakamoto K, Yamada H (1990) Biotechnol Lett 12: 357
105. Shimizu S, Kataoka M, Morishita A, Katoh M, Morikawa T, Miyoshi T, Yamada H (1990) Biotechnol Lett 12: 593
106. Hata H, Shimizu S, Hattori S, Yamada H (1989) Biochim Biophys Acta 990: 175
107. Hata H, Shimizu S, Hattori S, Yamada H (1990) J Org Chem 55: 4377
108. Shimizu S, Hattori S, Hata H, Yamada H (1988) Eur J Biochem 174: 37
109. Shimizu S, Kataoka M, Chung MCM, Yamada H (1988) J Biol Chem 263: 12077
110. Kataoka M, Doi Y, Sim TS, Shimizu S, Yamada H (1992) Arch Biochem Biophys 294: 469
111. Yamada H, Shimizu S, Kataoka M, Sakai H, Miyoshi T (1990) FEMS Microbiol Lett 70: 45

112. Kataoka M, Sakai H, Morikawa T, Katoh M, Miyoshi T, Shimizu S, Yamada H (1992) Biochim Biophys Acta 1122: 57
113. Kataoka M, Shimizu S, Yamada H (1992) Arch Microbiol 157: 279
114. Kita K, Matsuzaki K, Hashimoto, Yanase H, Kato N, Chung MCM, Kataoka M, Shimizu S (1996) Appl Environ Microbiol 62: 2303
115. Shimizu S, Yamada H, Hata H, Morishita T, Akutsu S, Kawamura M (1987) Agric Biol Chem 51: 289
116. Hata H, Shimizu S, Yamada H (1987) Agric Biol Chem 51: 3011
117. Sakamoto K, Kita S, Morikawa T, Shimizu S, Yamada H (1989) Japanese patent application H1-45407
118. Shimizu S, Sakamoto K, Yamada (1987) Nippon Nogeikagaku Kaishi 62: 283
119. Zhou B, Gopalan AS, Van Middlesworth F, Shieh WR (1983) J Am Chem Soc 105: 5925
120. Shimizu S, Kataoka M, Katoh M, Morikawa T, Miyoshi T, Yamada H (1990) Appl Environ Microbiol 56: 2374
121. Winkelman J, Lehninger AL (1958) J Biol Chem 233: 794
122. Yamada K (1959) J Biochem 46: 361
123. Bublitz C, Lehninger AL (1961) Biochim Biophys Acta 47: 288
124. Yamada K, Ishikawa S, Shimazono N (1959) Biochim Biophys Acta 32: 253
125. Kawada M, Takiguchi H, Kagawa Y, Suzuki K, Shimazono N (1962) J Biochem 51: 405
126. Hucho F, Wallenfels K (1972) Biochim Biophys Acta 276: 176
127. Weinberg R, Doudoroff M (1955) J Biol Chem 217: 607
128. Palleroni NJ, Doudoroff M (1957) J Bacteriol 74: 180
129. Rigo LU, Marechal LR, Vieira MM, Veiga LA (1985) Can J Microbiol 31: 817
130. Buchert J, Viikari L (1988) Appl Microbiol Biotechnol 27: 333
131. Kosuge T, Conn EE (1962) J Biol Chem 237: 1653
132. Fishbein WN, Bessman SP (1966) J Biol Chem 241: 4835
133. Fishbein WN, Bessman SP (1966) J Biol Chem 241: 4842
134. Shimizu S, Yamada H (1989) in: Vandamme EJ (ed) Biotechnology of Vitamins, Pigments and Growth Factors, Elsevier Applied Science, London, p. 199
135. Shimizu S, Yamada H (1990) in: Baldwin TO, Raushel FM, Scott AI (eds) Chemical Aspects of Enzyme Biotechnology, Plenum Press, New York, p. 151
136. Shimizu S, Kataoka M, Shimizu K, Hirakata M, Sakamoto K, Yamada H (1992) Eur J Biochem 209: 383
137. Kataoka M, Nomura J, Nose K, Shimizu S (1996) Nippon Nogeikagaku Kaishi 70: 374
138. Kataoka M, Shimizu K, Sakamoto K, Yamada H, Shimizu S (1995) Appl Microbiol Biotechnol 43: 974
139. Kataoka M, Shimizu K, Sakamoto K, Yamada H, Shimizu S (1995) Appl Microbiol Biotechnol 44: 333
140. Kataoka M, Hirakata M, Sakamoto K, Yamada H, Shimizu S (1996) Enzyme Microb Technol 19: 307
141. Bevinakatti HS, Newadkar RV (1989) Biotechnol Lett 11: 785
142. Glänzer BI, Faber K, Griengl H (1988) Enzyme Microb Technol 10: 689
143. Bergstrom S, Danielsson H, Samuelsson B (1964) Biochim Biophys Acta 90: 207
144. Van Dorp DA, Beerthuis RK, Nugteren DH, Vonkeman H (1964) Biochim Biophys Acta 90: 204
145. Jakschik BA, Sams AR, Sprecher H, Needleman P (1980) Prostaglandins 20: 401
146. Herold PM, Kinsella JK (1986) Am J Clin Nutr 43: 566
147. Oxdale L (1990) Chem Br 21: 813
148. Amano N, Shinmen Y, Akimoto K, Kawashima H, Amachi T, Shimizu S, Yamada H (1992) Mycotaxon 44: 257
149. Yamada H, Shimizu S, Shinmen Y (1987) Agric Biol Chem 51: 785
150. Shinmen Y, Shimizu S, Akimoto K, Kawashima H, Yamada H (1989) Appl Microbiol Biotechnol 31: 11
151. Shimizu S, Kawashima H, Akimoto K, Shinmen Y, Yamada H (1989) J Am Oil Chem Soc 66: 342
152. Shimizu S, Kawashima H, Akimoto K, Shinmen Y, Yamada H (1991) J Am Oil Chem Soc 68: 254
153. Shimizu S, Jareonkitmongkol S, Kawashima H, Akimoto K, Yamada H (1991) Arch Microbiol 156: 163

154. Jareonkitmongkol S, Shimizu S, Yamada H (1992) J Gen Microbiol 138: 997
155. Jareonkitmongkol S, Sakuradani E, Kawashima H, Shimizu S (1993) Nippon Nogeikagakukaishi 67: 531
156. Jareonkitmongkol S, Kawashima H, Shirasaka N, Shimizu S, Yamada H (1992) Appl Environ Microbiol 58: 2196
157. Jareonkitmongkol S, Sakuradani E, Shimizu S (1993) Appl Environ Microbiol 59: 4300
158. Jareonkitmongkol S, Shimizu S, Yamada H (1993) Biochim Biophys Acta 1167: 137
159. Jareonkitmongkol S, Sakuradani E, Shimizu S (1994) Arch Microbiol 161: 316
160. Kamada N, Jareonkitmongkol S, Kawashima H, Nishihara M, Shimizu S (1995) Nippon Nogei Kagaku Kaishi 69: 311
161. Jareonkitmongkol S, Kawashima H, Shimizu S, Yamada H (1992) J Am Oil Chem Soc 69: 939
162. Jareonkitmongkol S, Shimizu S, Yamada H (1993) J Am Oil Chem Soc 70: 119
163. Kawashima H, Kamada N, Sakuradani E, Jareonkitmongkol S, Shimizu S (1997) J Am Oil Chem Soc 73: in press

Glycobiotechnology: Enzymes for the Synthesis of Nucleotide Sugars

Lothar Elling
Institut für Enzymtechnologie, der Heinrich-Heine-Universität Düsseldorf im
Forschungszentrum Jülich, Postfach 2050, D-52404 Jülich, Germany

Dedicated to Professor Dr. Maria-Regina Kula on the occasion of her 60th birthday

Complex carbohydrates, as constituting part of glycoconjugates such as glycoproteins, glycolipids, hormones, antibiotics and other secondary metabolites, play an active role in inter- and intracellular communication. The aim of "glycobiotechnology" as an upcoming interdisciplinary research field is to develop highly efficient synthesis strategies, including in vivo and in vitro approaches, in order to bring such complex molecules into analytical and therapeutic studies. The enzymatic synthesis of glycosidic bonds by Leloir-glycosyltransferases is an efficient strategy for obtaining saccharides with absolute stereo- and regioselectivity in high yields and under mild conditions. There are, however, two obstacles hindering the realization of this process on a biotechnological scale, namely the production of recombinant Leloir-glycosyltransferases and the availability of enzymes for the synthesis of nucleotide sugars (the glycosyltransferase donor substrates). The present review surveys some synthetic targets which have attracted the interest of glycobiologists as well as recombinant expression systems which give Leloir-glycosyltransferase activities in the mU and U range. The main part summarizes publications concerned with the complex pathways of primary and secondary nucleotide sugars and the availability and use of these enzymes for synthesis applications. In this context, a survey of our work will demonstrate how enzymes from different sources and pathways can be combined for the synthesis of nucleotide deoxysugars and oligosaccharides.

List of Symbols and Abbreviations

A	Adenine
ADP	Adenosine 5'-diphosphate
ADP-Glc	Adenosine 5'-diphosphate α-D-glucose
C	Cytosine
CDP	Cytidine 5'-diphosphate
CDP-Glc	Cytidine 5'-diphosphate α-D-glucose
Cer	Ceramide
Fruc	D-fructose
Fuc	L-fucose
FucT	Fucosyltransferase
Gal	D-galactose
GalNAc	N-acetyl-D-galactosamine
GalNAcT	N-acetyl-D-galactosaminyltransferase
GalNH$_2$-1-P	Galactosamine-1-phosphate
GalT	Galactosyltransferase
GDP	Guanosine 5'-diphosphate
GDP-Fuc	Guanosine 5'-diphosphate β-L-fucose
GDP-Man	Guanosine 5'-diphosphate α-D-mannose
Glc-1-P	Glucose-1-phosphate
Glc-6-P	Glucose-6-phosphate
GlcNAc	N-acetyl-D-glucosamine
GlcNAcT	N-acetyl-D-glucosaminyltransferase
in situ Reg.	In situ regeneration
LacNAc	N-acetyllactosamine
Le	Lewis blood group
Man	D-mannose
ManT	Mannosyltransferase
NAD	Nicotinamide adenine dinucleotide
NADH	Nicotinamide adenine dinucleotide hydride
Nat.	Native
NDP	Nucleoside 5'-diphosphate
NeuAc	N-acetyl neuraminic acid
NMP	Nucleoside 5'-monophosphate
NTP	Nucleoside 5'-triphosphate
PEP	Phosph(enol)pyruvate
PP	Pyrophosphorylase
Pyr	Pyruvate
Rec.	Recombinant
Ser	L-serine
SuSy	Sucrose synthase
Thr	L-threonine
dTDP	2'-deoxythymidine 5'-diphosphate

dTDP-Glc	2'-deoxythymidine 5'-diphosphate α-D-glucose
U	Uracil
dU	2'- deoxy-uridine
UDP	Uridine 5'-diphosphate
UDP-Ara	Uridine 5'-diphosphate β- L-arabinose
UDP-Gal	Uridine 5'-diphosphate α-D-galactose
UDP-GalNAc	Uridine 5'-diphosphate N-acetyl-α-D-galactosamine
UDP-GalNH$_2$	Uridine 5'-diphosphate α-D-galactosamine
UDP-Glc	Uridine 5'-diphosphate-α-D-glucose
dUDP-Glc	2'-deoxy-uridine 5'-diphosphate α-D-glucose
UDP-GlcA	Uridine 5'-diphosphate α-D-glucuronic acid
UDP-GlcNAc	Uridine 5'-diphosphate N-acetyl-α-D-glucosamine
UDP-Xyl	Uridine 5'-diphosphate α-D-xylose
UMP	Uridine 5'-monophosphate

1 Introduction

In the last two decades "Glycobiotechnology" has emerged as a new inter-disciplinary research field. Glycobiotechnology includes many research areas, spanning from basic sciences to applied sciences and medicine. Molecular cell biology and glycobiology are also incorporated, as well as enzyme technology and biotechnology, synthetic organic and bioorganic chemistry, and pharmocology. The general task of glycobiotechnology is the transfer of the basic knowledge about structure-function relationships of glycoconju-gates to practice-related synthetic and applied procedures. The term glyco-conjugates originally comprises glycoproteins and glycolipids and has been created by Montreuil [1] in 1973. By extension many other glycosylated struc-tures are now included in this term. Montreuil [1] points out by two quotations what can be expected from glycobiotechnology in basic and applied research areas:

"Together with nucleic acids and proteins, carbohydrates represent the third dimension of molecular biology" (Francois Jacob).

"A spoonful of sugar makes dividends go up. People say that in five years, there will be more companies with glyco in their name than companies with gene in their name today" (Herald Tribune August 22, 1990).

However, the latter prediction is still to be awaited, since glycobiology has still to reveal more functions of the microheterogeneous oligosaccharrides on proteins, lipids and other aglycones.

Complex carbohydrates, as constituting part of glycoconjugates such as glycoproteins, glycolipids, hormones and antibiotics, play an active role in inter- and intracellular communication. It has been shown that oligosaccharides mediate specific recognition events and provide modulation of biological pro-cesses. Oligosaccharides are regarded as multifunctional "information carriers" and can act as specific receptors in host-pathogen (e.g. viruses, bacteria and parasites) interactions [2–10], as ligands for cell adhesion molecules in cell-cell interactions (e.g. inflammation, tumor development and metastasis, blood groups) [11–17] or as modulators of biological active compounds (e.g. anti-biotics, hormones and enzymes) [18–24]. It is beyond the scope of this review to cover all the literature describing the biosyntheses, role and function of glycoconjugates. The reader is referred to recently published excellent reviews covering these topics [25–33].

To bring such complex molecules or mimetica thereof into analytical and therapeutic studies highly efficient strategies for their syntheses are required. A great challenge still lies in the synthesis of the "glyco"-part of glycoconjugates. The variety of glycoconjugates is due to the fact, that in oligosaccharides the specific sequence of monosaccharides in their different conformations and the regio- and stereospecific formation of the glycosidic bond as well as the branch-ing pattern is different. In theory the formation of a disaccharide consisting of two different monosaccharides gives rise to 20 isomeric structures [34]. This illustrates that each step in the synthesis of oligosaccharides has to meet the requirements for high regio- and stereoselectivity.

Many efficient chemical synthesis procedures have been developed [35–38]. However, for the synthesis of each particular oligosaccharide structure an individual strategy has to be set up [35]. Therefore, the chemical synthesis of oligosaccharides comprises multiple steps with protection and deprotection of the functional groups for stereo-controlled synthesis resulting often in time-consuming steps with only moderate product yields. In general, the coupling of one monosaccharide component comprises 5 steps with one week of lab work for each step [39], e.g. the synthesis of N-acetyllactosamine (LacNAc) consists of 12 steps and needs 3 months of lab work [39]. Since almost all steps are carried out in organic solvents, environmental problems arise when a scale-up of the chemical procedure is envisaged.

On the other hand, the enzymatic synthesis of glycoconjugates and oligosaccharides leads to high product yields in a short time by stereo- and regioselective one-step reactions. All enzymatic reactions are easy to scale up and are carried out in aqueous media under mild conditions. A whole set of enzymes is now available to build up C–C bonds in monosaccharides, C–O–P bonds in activated monosaccharides e.g. phosphorylated sugars or nucleotide sugars, and C–O–C bonds in di- and oligosaccharides (Fig. 1). However, all these enzymatic reactions are limited by the substrate spectrum of the individual enzyme.

Since neither the chemical nor the enzymatic synthesis alone can assist in the rational design of a new generation of therapeutic carbohydrate target structures, the chemoenzymatic approaches have been successfully utilized [40–42].

In the following sections some examples of glycoconjugates which have attracted the attention of glycobiologists will be discussed, together with a short reflection on the availability of Leloir-glycosyltransferases for glycobiotechnological processes. Finally, the main objective of this review is to give an insight into the complex pathways of primary and secondary nucleotide sugars and

Enzymes for Carbohydrate Synthesis		
Bond C - C	C - O - P	C - O - C
Enzymes Aldolases Transketolases	Pyrophospho - rylases	Glycosyl - transferases Glycosidases
Products Mono - saccharides	Nucleotide Sugars	Di- and Oligo- saccharides

Fig. 1. Enzymes for the synthesis of carbohydrates

their importance for glycobiotechnological applications. In this respect, an overview about our approaches is presented in order to demonstrate how enzymes from different pathways can be combined with the appropriate reaction technology to obtain activated (deoxy)sugar as donor substrates for glycosyl-transferases.

2 Glycoconjugates as Synthetic Targets

2.1 Eucaryotic Glycoconjugates

The terminal oligosaccharide structures (glycans) in glycoconjugates which are associated with the eucaryotic cell represent specific contact and recognition sites for other cells, microorganisms, viruses, proteins, hormones and toxins.

In eucaryotic cells the glycans on proteins are covalently linked to the amide group of an asparagine (N-glycans) or to a hydroxyl group of a serine or threonine (O-glycan) [25]. Typical features of N-glycans are (i) the specific glycosylation motif Asn-X-Ser/Thr in proteins; (ii) the common core structure Man(α1–3) [Man(α1–6)] Man(β1–4) GlcNAc(β1–4) GlcNAcβ-Asn; (iii) the classification in the oligomannosidic, complex or hybrid type; (iv) the cotranslational en bloc transfer of an oligosaccharide precursor from the membrane-bound lipid carrier dolichol in the rough endoplasmatic reticulum; and (v) the trimming, branching, elongation and termination in the golgi apparatus. In complex-type N-glycans, multi-branched (antennary) structures are built up by up to six different GlcNAc glycosyltransferases (Fig. 2) and define the multivalent presentation of cell-specific terminal structures on the cell surface [31,43–45]. The different branches of complex N-glycans carry one or more N-acetyllactosamine (LacNAc or Poly-LacNAc) building blocks to which L-fucose (Fuc), N-acetyl-neuraminic acid (NeuAc), N-acetyl-D-galactosamine (GalNAc) or D-galactose (Gal) is attached.

This is also valid for O-glycans of the mucin type and glycolipids [46–48]. O-glycans of the mucin type are, however, less branched than complex N-glycans, but they can reach a high cluster-like density of terminal oligosaccharide structures by the appearance of multiple O-glycosylation sites (tandem repeats) in the polypeptide backbone [43,46,49]. A further typical feature of O-glycans is that the biosynthesis proceeds by the sequential linkage of monosaccharides in the smooth endoplasmatic reticulum and the Golgi apparatus [50]. The structures of O-glycans with their common Tn-antigen (GalNAcα-Ser/Thr) are classified in 7 core structures [46]. Poly-LacNAc structures are mainly found as extensions of the core 2 structure (Fig. 3). All core structures carry Fuc, NeuAc, Gal or GalNAc at their non-reducing termini.

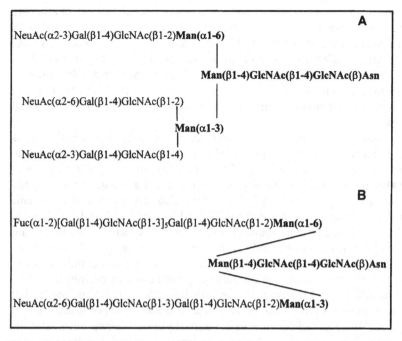

Fig. 2. *N*-glycans of the hybrid type. A: triantennary structure with sialylated *N*-acetyllactosamine (LacNAc) building blocks; B: biantennary structure with sialylated or fucosylated poly-LacNAc building blocks. The *N*-glycan core structure is written in bold letters

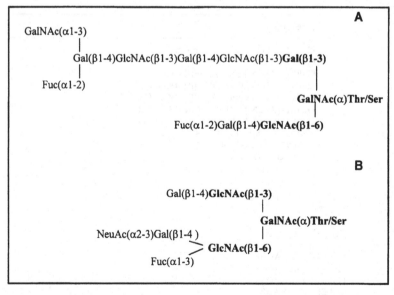

Fig. 3. *O*-glycans of the mucin type. A: "core 2" structure with poly-LacNAc and blood group epitope A; B: "core 4" structure with sialyl Le^x epitope. The core structures are written in bold letters

Glycosphingolipids as important components of the cell membrane are constituted of a ceramide (Cer, sphingosine and fatty acid) moiety to which an oligosaccharide chain is covalently linked. According to the structure of the oligosaccharides different structures can be identified (Table 1) [51–54]. These are synthesized by the sequential transfer of monosaccharides by different glycosyltransferases in the Golgi apparatus [15, 27, 47]. Gal, GalNAc, Fuc and NeuAc appear at the non-reducing end as residues of cell-type-specific terminal structures.

In summary, disaccharides in glycoproteins and glycolipids can be defined as carriers of terminal structures [47] (Table 2). The most important terminal structures are summarized in Fig. 4. They comprise antigenic determinants of the AB0(H) and Lewis (Le) blood groups, the i and I structure as well as the T and Tn antigen [46, 47]. The Le^x structure also represents a stage-specific antigen (SSEA) and a tumor-specific marker [15, 16, 55]. Le^x and Le^a serve in their α2-3 or sulfated form as ligands for E-, P- and L-selectins [28]. E-, P- and L-selectins belong to a new family of mammalian adhesion molecules which initiates in a Ca^{2+}-dependent manner the primary events of the inflammatory cascade and the recruitment of circulating lymphocytes at peripheral lymph nodes. The adhesion of leucocytes to endothelial cells and to high endothelial venules in peripheral lymph nodes as well as the metastasis of tumor cells is mediated by the interaction of specific carbohydrate ligands with selectins [11, 13, 43, 56–59]. Pathological disorders concerning the inflammatory cascade

Table 1. Different classes of glycosphingolipid core carbohydrate structures (taken from [54] with kind permission)

Class	core structure
Arthro	GlcNAc(β1-3)Man(β1-4)Glc-Cer
Gala	Gal(α1-4)Gal-Cer
Ganglio	Gal(β1-3)GalNAc(β1-4)Gal(β1-4)Glc-Cer
Globo	GalNAc(β1-3)Gal(α1-3)Gal(β1-4)Glc-Cer
Globoiso	GalNAc(β1-3)Gal(α1-3)Gal(β1-4)Glc-Cer
Lacto	Gal(β1-3)GlcNAc(β1-3)Gal(β1-4)Glc-Cer
Lactoneo	Gal(β1-4)GlcNAc(β1-3)Gal(β1-4)Glc-Cer
Muco	Gal(β1-3)Gal(β1-4)Gal(β1-4)Glc-Cer
Mollu	Man(α1-3)Man(β1-4)Glc-Cer
Schisto	GalNAc(β1-4)Glc-Cer

Table 2. Peripheral disaccharides carrying terminal structures according to [47]

Type	Peripheral Disaccharides
Type 1	Gal(β1-3)GlcNAc(β)
Type 2	Gal(β1-4)GlcNAc(β)
Type 3	Gal(β1-3)GalNAc(α)
Type 4	Gal(β1-3)GalNAc(β)
Type 6	Gal(β1-4)Glc

include chronic diseases like psoriasis and rheumatoid arthritis as well as diseases like acute inflammation of the lung, the digestive system, pancreas and myocard [11, 13]. These examples demonstrate the great therapeutic potential of synthesized oligosaccharides as anti-adhesiva [13, 60].

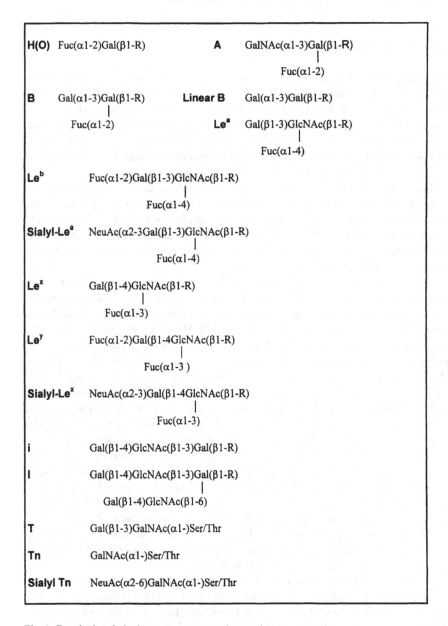

Fig. 4. Terminal carbohydrate structures on glycoconjugates

The Gal(α1-3) Gal epitop is found on cell surfaces of many mammalians and New World monkeys, but not in humans, primates or Old World monkeys. In human blood up to 1% of the circulating IgG molecules are directed against this epitop [30, 44] presenting an enormous immunological barrier for xenotransplantation of animal organs (e.g. from pig) to humans.

2.2 Procaryotic Glycoconjugates

The dogma that glycoproteins occur exclusively in eucaryotic cells had to be revised in the last twenty years. Bacterial glycoproteins form the outer surface (S)-layer of archae- and eubacteria [61, 62]. In analogy to eucaryotic cells, partly sulfated oligosaccharides attached to Asn or Thr are found in halobacteria. Nucleoside-diphosphate-activated oligosaccharides were identified as intermediates in the biosynthesis of the S-layer of the eubacterium *Bacillus alvei* [63] and the archaebacterium *Methanothermus fervidus* [64].

Nucleoside diphosphate disaccharides and peptides are the anabolic precursors of pseudomurein, the peptidoglycan of archaebacteria [65, 66]. The central intermediate is a UDP-activated glycopentapeptide consisting of a pentapeptide which is covalently linked to the disaccharide *N*-acetyltalosaminuronic acid (β1-3)*N*-acetylglucosaminide. In eubacteria a pentapeptide is attached to UDP-activated *N*-acetylmuraminic acid (**1**) (Fig. 5).

Endotoxins comprise the lipopolysaccharides (LPS) or enterobacteria and interact with macrophages of the host (human) causing in high doses unspecific acute responses by an excessive release of endogeneous mediators (cytokines, prostaglandines, leukotrienes and others) [67]. However, in small doses endotoxins produce beneficial effects. Lipopolysacchardes affect the differentiation, proliferation and cytotoxicity of lymphocytes against tumors. The endotoxic effective component in the structure of lipopolysaccharides is the lipid A. It consists of a 1,4'-bisphosphorylated β-D-glucosaminyl-(1-6)-α-D-glucosaminide to which four mole of (R)-3-hydroxy fatty acids are attached in the C2- and C3-position (**2**) (Fig. 6) [67, 68]. Additional important components of enterobacteria are the O-specific antigens in LPS and the cell-associated capsular polysaccharides. Both determine the serological specificity of enterobacteria. They consist of an oligosaccharide repeating unit which is substituted by other monosaccharides in a species- and serotype-dependent manner. In the O-antigen, Man, Gal, Glc and GlcNAc appear besides 6-deoxy and 3, 6 dideoxysugars whose biosyntheses pathway are currently being investigated [4, 5, 19, 69]. In summary, the synthesis of lipid A or the development of antibodies against lipid A or O-specific oligosaccharides raise the prospects of finding suitable therapeutica for the treatment of the sepsis syndrome.

An interesting aspect of enterobacterial O-antigens is the occurence of mammalian oligosaccharide structures. Among those ABO(H) blood groups, the H-type I (Fuc(α1-2) Gal(β1-3)GlcNAc) and the Le^x structure can be found [70–72]. The latter is found in the O-antigen of *Helicobacter pylori* strains,

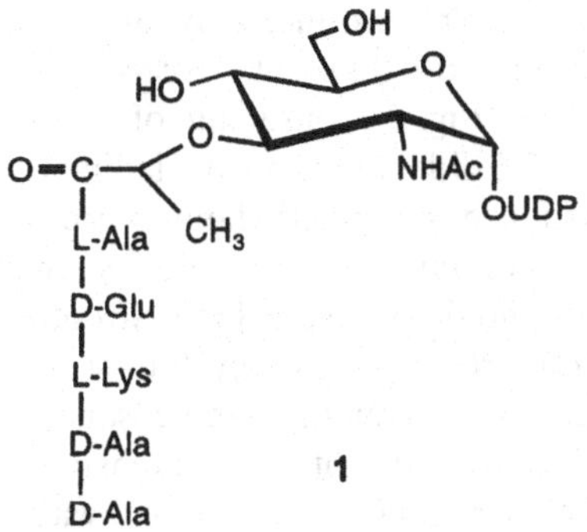

Fig. 5. UDP-*N*-acetyl-muraminic acid pentapeptide (**1**), an intermediate in the biosynthesis of murein in eubacteria

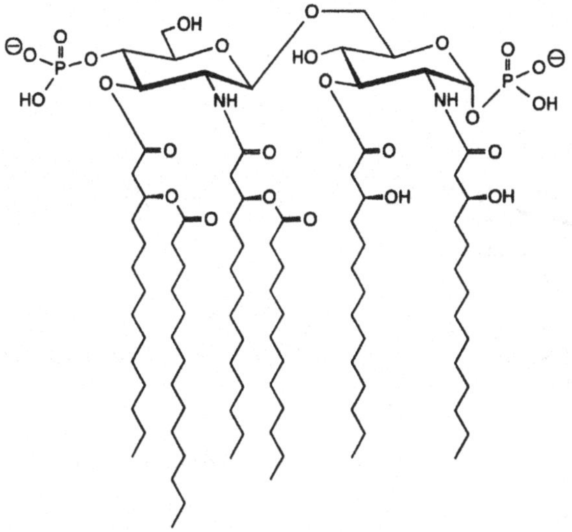

Fig. 6. Example of a lipid A (**2**) molecule (according to [68])

a bacterium which has been implicated as a causative agent of gastritis, gastric and duodenal ulcers, and gastric carcinoma. It is suggested that the bacterium escapes from the recognition by the host immune system by presenting host-specific antigens as a molecular mimicry mechanism. The enzymes for the synthesis of Le^x in *Helicobacter pylori* have been recently characterized [73].

The classical definition of glycoconjugates can also be extended to compounds where the glyco-part is connected to structural components other than proteins or lipids. Secondary metabolites of plants and microorganisms, e.g. digitalis glycosides and antibiotics belong to these glycoconjugates. They have in common that unusual deoxysugars contribute to their efficacy and pharmacological kinetics. The macrolides which are produced by different actinomycetes contain macrocyclic lactone rings and different deoxysugars. A well known example is erythromycin A (3) (Fig. 7) consisting of a 14-membered lactone ring which is substituted by a 2,6 dideoxyhexose (L-cladinose) and a 3,4,6 trideoxyhexose (D-desosamine) with a basic dimethylamino group [74]. Streptomycin is an example from another family of antibiotics, the aminoglycosides, which are built up by cyclitols and deoxysugars [75]. Streptomycin (4) represents a pseudotrisaccharide with streptidine as a cyclitol structure and the deoxysugars dihydrostreptose and N-methyl-L-glucosamine (Fig. 8). The molecular genetics of the biosynthesis of antibiotics are intensively investigated [19,76,77] in order to make the metabolic design of pathways possible in the near future [18].

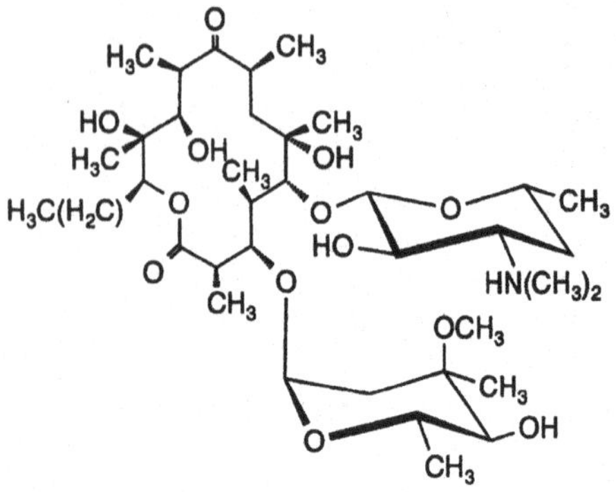

3

Fig. 7. Structure of erythromycin A (3)

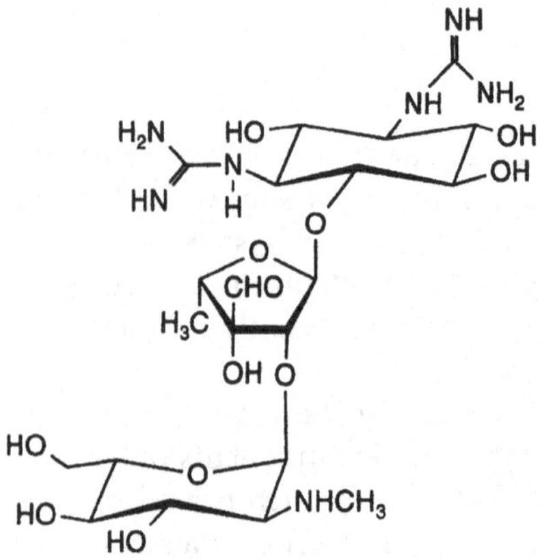

4

Fig. 8. Structure of streptomycin (4)

3 Enzymes for the Synthesis of Glycoconjugates

3.1 Glycosidases

Glycosidases catalyze in vivo the hydrolysis of glycosidic bonds with retention (β-galactosidase and lysozyme) or inversion (trehalase and β-amylase) of the anomeric configuration [78]. The mechanism of the enzyme-catalyzed hydrolysis resembles the acid-catalyzed chemical hydrolysis of glycosides. In both cases an oxonium ion or a transition state with the character of an oxonium ion exists. A carboxylate function of the glycosidases stabilizes the intermediate or the transition state [78]. However, in most cases evidence is lacking whether an oxocarbenium ion as a stabilized ion pair or a glycosyl ester exists. It has been shown for a β-glucosidase from *Agrobacterium* that a carboxylate as a nucleophile cleaves the glycosidic bond and forms an enzyme-glycosyl ester. In a further reaction a second nucleophile, H_2O or the hydroxyl group of an alcohol or monosaccharide, cleaves the enzyme-glycosyl ester [79–81].

Glycosidases are utilized for the synthesis of glycosides under thermodynamically or kinetically controlled reaction conditions (Fig. 9) [82]. The thermodynamic approach is the reverse hydrolysis. Since the thermodynamic equilibrium prefers hydrolysis, suitable reaction conditions like high substrate concentrations, high temperature or the addition of organic solvents have to be chosen to shift the equilibrium to the product site [83–86]. In kinetically controlled syntheses (tranglycosylation), activated glycosyl donors (8–10) are used. Their reactive intermediates are captured by stronger nucleophiles, e.g. H_2O or the hydroxyl group of an alcohol or a monosaccharide [82]. Activated

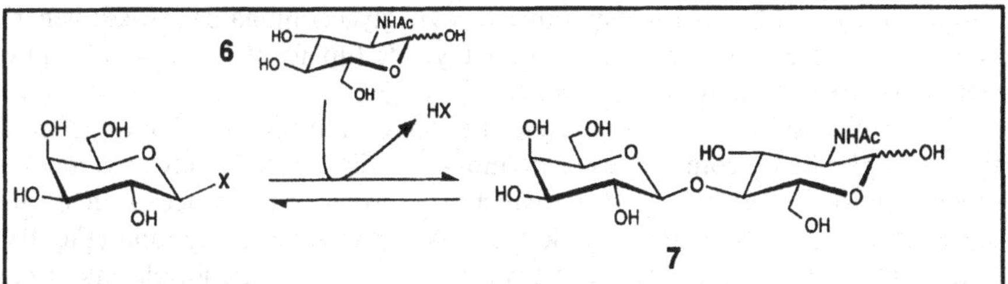

Thermodynamic: X = OH (5)

Kinetic: X = F (8), o-NO$_2$C$_6$H$_4$ (9), OR (R = Glc) (10)

Fig. 9. Synthesis of *N*-acetyllactosamin (LacNAc, **7**) with β1-4 galactosidase. The thermodynamic approach (reverse hydrolysis) involves D-galactose (**5**) as donor and *N*-acetyl-D-glucosanmine (**6**) as acceptor substrate. Activated galactosyl donors (**8–10**) are used in the kinetic approach (transglycosylation)

glycosyl donors are glycosylfluorides, aromatic glycosides or di- and oligosaccharides and their consumption during the synthesis reaction must be carefully
followed because hydrolysis is still thermodynamically favoured [87].

Glycosidases give relatively low product yields between 10% and 40% in the
reverse hydrolysis or transglycosylation reaction. Isolation of the product is
complicated by the relatively high donor and acceptor concentrations. Some
glycosidases lack absolute regioselectivity, which leads in the case of β-galactosidase to the formation of allo-N-acetyllactosamine (Gal(β1-6GlcNAc) besides
the main product N-acetyllactosamine [88–90] (Fig. 9). These regioisomers have
to be separated from the main product by laborious and expensive chromatographic procedures [91]. However, regioselectivity can be directed and controlled by the anomeric configuration of the acceptor; α1-3 and α1-6 glycosides were
preferentially formed with α-galactosidase by the combination of Gal-O-p-
$NO_2C_6H_4$ with α-Gal-O-Me and β-Gal-O-Me, respectively [84,92]. The favoured formation of one regioisomer by the use of α-anomeric acceptors has
also been shown for β-N-acetylhexosaminidase [93–95].

However, a summary of structures synthesized with glycosidases [96,97]
reveals that some biologically relevant structures occuring in glycoconjugates
can be synthesized by these enzymes, but most of them are only disaccharide
structures. Therefore, glycosidases alone are not suitable for building up the
diversity of oligosaccharide structures of glycoconjugates in vitro. In summary,
the application of these budget-priced enzymes discloses basic disadvantages
with reference to the efficiency and economy of the synthesis as well as the purity
of the products.

3.2 Leloir Glycosyltransferases

The application of Leloir glycosyltransferases in glycoconjugate synthesis generally offers two advantages: high product yields (up to 100%) and absolute
stereo- and regioselectivity.

Leloir glycosyltransferases catalyze the stereo- and regiospecific transfer of
the glycosyl moiety from a nucleoside mono- or diphosphate-activated monosaccharide to an acceptor substrate which is a mono- or oligosaccharide or an
aglycon such as a protein, a peptide, a lipid or another compound (Fig. 10)
[31,98,99]. Leloir glycosyltransferases are localized in the endoplasmatic reticulum and Golgi apparatus as transmembrane type II glycoproteins (N-termi-
nus inside/C-terminus outside). This feature renders their isolation from tissues
more difficult and detergents have to be used [100,101]. Over 100 cDNAs of
eucaryotic glycosyltransferases have been cloned up to now [100,102–104], and
yet there are only a few homologues among them, all of them are characterized
by a common domain structure. They consist of a short N-terminal segment,
a transmembrane non-cleavable signal/anchor sequence, a proteolytic cleavable
stem region and a C-terminal globular catalytic domain [44,100,105]. Published data for β1-4 galactosyltransferase demonstrate that the transmembrane

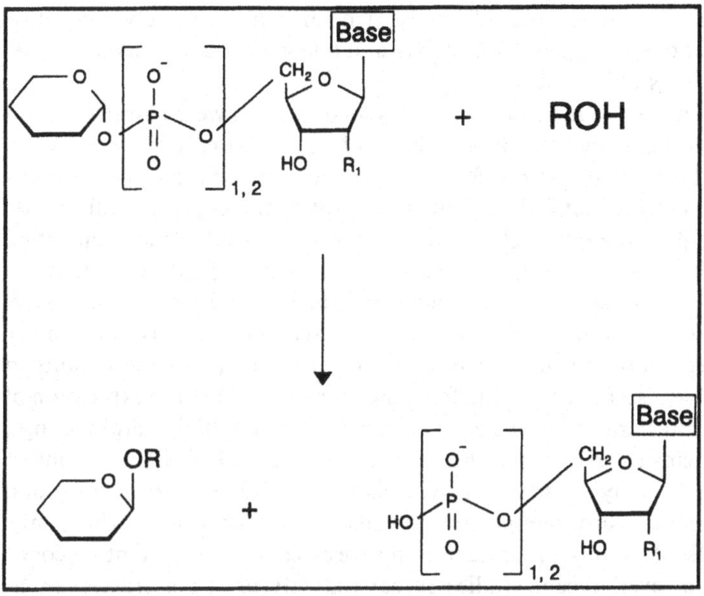

Fig. 10. General reaction scheme of Leloir-glycosyltransferases. Base = $U(R_1 = OH)$, $T(R_1 = H)$, $C(R_1 = OH)$, $A(R_1 = OH)$, $G(R_1 = OH)$ and R = protein, lipid or aglycon

domain contains a specific sequence for the retention of the enzyme in the Golgi apparatus and that the cytoplasmic part plays a non-specific supporting role [106–110]. The transmembrane domain and flanking charged residues are necessary and sufficient for the localization of the $\alpha 2$-6 sialyltransferase in the Golgi apparatus [44, 111]. Similar results were also obtained for $\beta 1$-2 N-acetyl-glucosaminyltransferase I [112, 113]. Publications describing the isolation of soluble and enzymatically active glycosyltransferases have shown that these enzymes can be cleaved in the stem region releasing the catalytic domain [99, 100, 101, 114–119]. It is still unclear whether this is due to an in vivo proteolytic cleavage or an artefact during enzyme purification [44]. However, an important fact is that the soluble catalytic domain is fully active and contains catalytically important amino acid residues. In $\beta 1$-4 galactosyltransferase from human milk the potential binding site for UDP-Gal and/or the acceptor substrate contains three Tyr and two Trp residues [120], two essential Cys residues forming a disulfide [121] as well as essential Phe, Pro, and Asn residues [122]. Two characteristic conserved sequence motifs with a Cys residue in each of them ("sialylmotif") are responsible for the binding of the donor substrate CMP-NeuAc in sialyltransferases [123–126]. Conserved Cys residues are also found in the catalytic domain of $\alpha 1$-3 fucosyltransferase III and V which are replaced by a Ser residue in $\alpha 1$-3 fucosyltransferase IV [127]. A Lys residues is probably involved in the binding of GDP-Fuc [128].

Such characteristic data are yet not known for procaryotic glycosyltransferases. Many of them transfer deoxysugars and a few will be soon available for biochemical studies [18, 19, 69, 129].

In the last few years, the synthesis of glycoconjugates were summarized in a number of excellent reviews [40, 42, 130–137]. A crucial requirement for the application of Leloir glycosyltransferases in preparative syntheses is the availability of these enzymes and their donor substrates, the expensive nucleotide sugars. According to the model of soluble enzymes, several genes encoding Leloir glycosyltransferases have been cloned and expressed in their truncated form without the *N*-terminal and transmembrane domain and also partly without their stem region. In combination with secretion signals and affinity tags, this strategy allows glycosyltransferases from cell-culture supernatants to be purified. Table 3 summarizes data from the literature where the expression of Leloir glycosyltransferases yielded enzyme activities in the 10^{-3} U and U range. Expression systems are *E. coli*, *Saccharomyces cerevisiae*, CHO, COS and insect cells. Surprisingly, only β1-4 galactosyltransferase, α2-3 sialyltransferase and α2-6 sialyltransferase are commercially available at high prices, which may be due to a lack of optimized fermentation processes of recombinant glycosyltransferases. However, recent studies reveal that glycosyltransferases can be produced on a large scale by fermentation (10 to 150 I) [138–140, 146].

Table 3. Recombinant mammalian Leloir-glycosyltransferases expressed in the mU to U scale[3]

Transferase/Source	Expression-System	Activity (U)[1]	Protein[2]	References
Galactosyltransferases (GalT)				
β1-4 GalT				
EC 2.4.1.38/90				
human	E. coli	$5 \times 10^{-4\,b**}$	cd, s,	[120]
human	E. coli	105^{a*}	cd, af,	[122]
human	E. coli	$14 \times 10^{-4\,a*}$	cds,	[141]
HeLa	S. cerevisiae	4.7^{a*}	fl	[142]
	S. cerevisiae	2.0^{a*}	cds, s	[140]
α1-3 GalT				
EC 2.4.1.124/151				
mouse	COS-1	$0.16 \times 10^{-3*}$	cd, s, af	[143]
Bovine	Insect cells	3^{a*}	fl	[144]
α1-3 GalT				
(blood group B)				
EC 2.4.1.37				
human	E. coli	1.36^{a*}	cds, s	[145]
Mannosyltransferases (ManT)				
α1-2 ManT				
EC 2.4.1.?				
Yeast	E. Coli	$0.166^{a,*}$	cd, s	[139, 146]

Table 3. (*Continued*)

Transferase/Source	Expression-System	Activity (U)[1]	Protein[2]	References
Sialyltransferases (SiaT)				
α2-6 SiaT (N-Glycans) EC 2.4.99.1 human	S. cerevisiae	3×10^{-4}[a][**]	fl	[138]
α2-6 SiaT (O-Glycans) EC 2.4.99.6 Porcine	Insect cells	3.1[a][*]	cd, s,	[147]
N-Acetylglucosaminyltransferases (GlcNAcT)				
β1-2 GlcNAcTI EC 2.4.1.101 Rabbit	Insect cells	—	—	[148]
Rabbit	Insect cells	2[a][*]	cd, s	[149]
N-Acetylgalactosaminyltransferases (GalNAcT)				
α1-3 GalNAcT (Mucin Type) EC 2.4.1.? human	Insect cells	0.75[a][*]	fl	[150]
Bovine	Insect cells	2.5[a][*]	cds, s	[151]
Fucosyltransferases (FucT)				
α1-3/4 FucT EC 2.4.1.65 human	COS-1	4×10^{-4}	cd, s, af	[152]
α1-3 FucT IV EC 2.4.1.152 Mouse	COS-7	3×10^{-4}[a][**]	fl	[153]
human	COS	17×10^{-5}[b][*]	cd, af	[154]
α1-3 FucTV EC 2.4.1.152 human	COS	7×10^{-4}[b][*]	cd, af	[154]

[1] $U = \mu mol\,min^{-1}$; [*] purified enzyme; [**] crude extract; [a] specific activity (Umg^{-1});
[b] volumetric activity ($U\,ml^{-1}$);
[2] af: affinity marker, cd: catalytic domain, cds catalytic domain + stem region, fl: full length gene, s: secretory
[3] see also note added in proof.

4 Primary and Secondary Nucleotide Sugars

Primary and secondary nucleotide sugars are needed for the preparative synthesis of glycoconjugates with glycosyltransferases. Enzymes for the formation

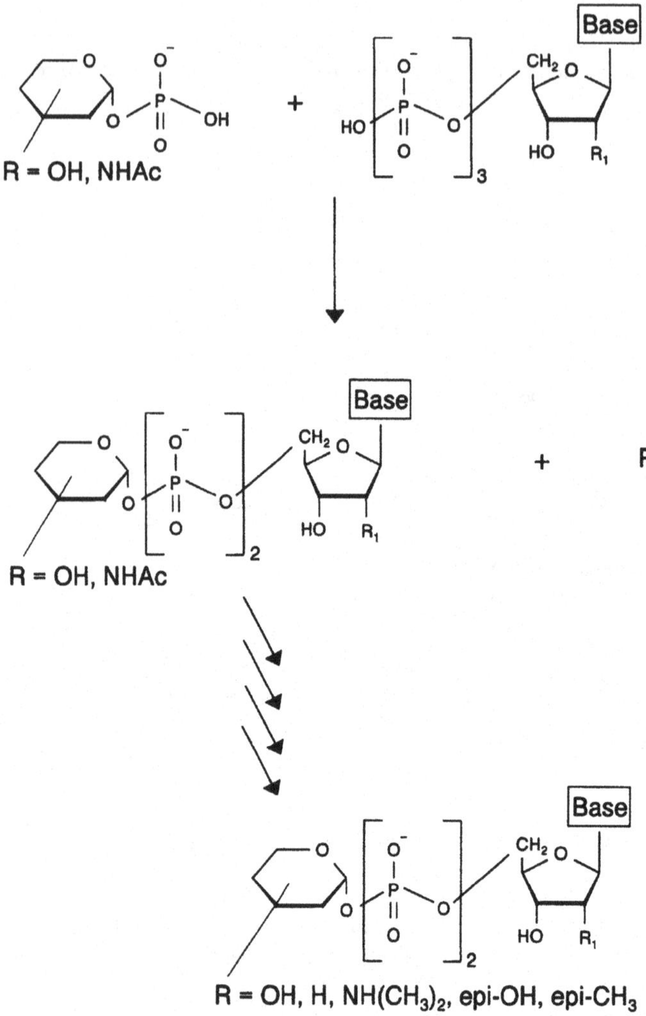

Fig. 11. General reaction scheme of the Leloir biosynthesis pathways for primary and secondary nucleotide sugars. Base = U (R_1 = OH), T (R_1 = H), C (R_1 = OH), A(R_1 = OH), G (R_1 = OH). The arrows symbolize multiple enzymatic transformations

of primary nucleotide sugars are pyrophosphorylases which convert a sugar-1-phosphate and a nucleoside triphosphate to the corresponding activated monosaccharide (Fig. 11) [155]. Secondary nucleotide sugars are generated by further enzymatic modifications in the sugar moiety of the primary nucleotide sugars, e.g. epimerisation, deoxygenation, amination and methylation [19, 156, 157]. We will see in the following chapters that many of the biochemical pathways for secondary nucleotide sugars still have to be elucidated and that nearly all the enzymes involved are still not available for large-scale preparative synthesis.

However, the alternative way to produce activated deoxysugars is by chemical synthesis, with all its formerly described disadvantages. In this review only a few examples will be mentioned. Chemical syntheses of the GDP-activated 6-deoxysugar L-fucose [158–161] and dTDP-activated 3,6 didesoxyglucose [162] have been worked out. Chemical methods are clearly in favour when derivatives of nucleotide sugars are envisaged for testing the substrate specificity of glycosyltransferases. In this context deoxysugar derivatives of UDP-Gal [163] and GDP-Fuc [164–167], biotin-labeled GDP-Fuc [168] and fluorescent-labelled CMP-NeuAc have been synthesized. In general, a sugar-1-phosphate is obtained which is coupled to a chemically activated nucleotide resulting in low yields [96, 169, 170].

4.1 Pathways of Nucleotide Deoxysugars

D- and L-deoxysugars are widespread as components of glycoconjugates [171, 172]. In mammalians L-fucose (6-deoxy-L-galactose, 11) is found in glycoproteins and glycolipids [173]. A great structural diversity of D- and L-deoxysugars is found in secondary metabolites of pro- and eucaryotes, e.g. antibiotics (macrolides, anthracylines, aminoglycosides), digitalis glycosides or other plant metabolites. These structures are synthesized in the nucleotide sugar pathways which are summarized in Figs. 12 and 13. The central intermediate for all deoxysugars is the nucleoside-diphosphate-activated 6-deoxy-4-ketohexose intermediate. The nucleobases involved are characteristic for the pathways in plants (uracil, U), in bacteria with reference to the synthesis of secondary metabolites, components of cell walls (deoxythymine) and certain sugars (cytosine) for bacterial O-antigen synthesis as well as in mammalian cells for sugar synthesis (guanine) of glycoconjugates (Fig. 12 and 13). The activated 6-deoxy-4-ketohexose is the branching point for D- and L-deoxysugar pathways. Pathways for the activated L-deoxysugars are entered by a 3,5 epimerisation and a variety of subsequent steps, e.g. isomerisation, oxidation, transamination and methylation leads to multiple deoxygenated, aminated and branched sugars. It is obvious that a whole enzyme cascade is necessary to build up D- and L-deoxysugars. The in vitro synthesis of nucleotide deoxysugars is a crucial prerequisite for the identification and biochemical studies of corresponding glycosyltransferases. Apart from mammalian fucosyltransferases (see Table 3) only a few deoxysugar transferring enzyme have been investigated [19, 12, 174, 175].

4.2 Salvage Pathways

Salvage pathways of nucleotide sugars represent biochemical "short-cuts" by which the cellular metabolism circumvents complex pathways for secondary nucleotide sugars. These de novo syntheses involve kinases phosphorylating the

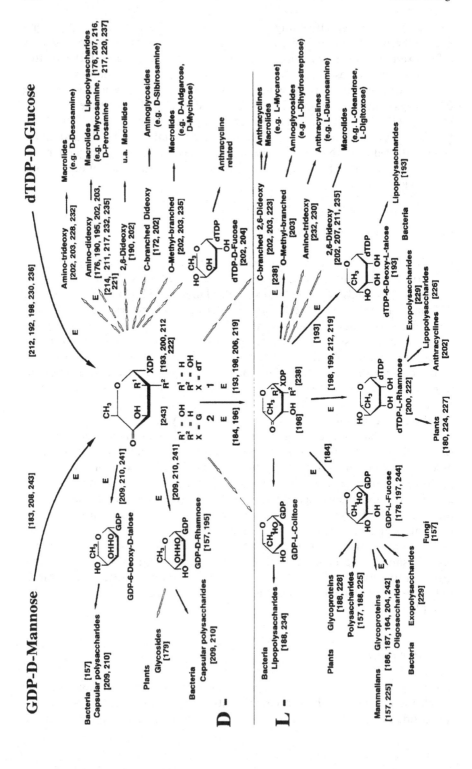

monosaccharide at the anomeric hydroxyl group (**12, 15, 18**) and specific pyrophosphorylases for activation with nucleoside triphosphates. Salvage pathways for GDP-Fuc (**13**), UDP-Gal (**16**) and UDP-GalNAc (**19**) (Fig. 14) have been mainly described in mammalian tissues [245–251] and plants [189, 252–254]. According to the above given definition (see Fig. 11), these secondary nucleotide sugars can then be classified as primary nucleotide sugars. The "direct activation" of L-fucose (**11**) (Fig. 14) has been already utilized for the synthesis of **13** [255] and the fucosylation of glycoconjugates in combination with fucosyltransferases [256].

5 Enzymes for the Synthesis of Primary and Secondary Nucleotide Sugars

There are two ways of producing nucleotide sugars by enzymatic syntheses: (i) the large-scale (gram scale) enzymatic syntheses [257] for stoichiometric reactions with glycosyltransferases where the inhibitory nucleoside diphosphate is cleaved by alkaline phosphatase [258] ; (ii) the in situ regeneration of nucleotide sugars [131, 256, 259] where only catalytic amounts of the nucleotide sugar are synthesized by a complex recycling enzyme system (see Sect. 7). Both approaches assume that the enzymes for the synthesis of primary and secondary nucleotide sugars are available in sufficient amounts and are well characterized with regard to their stability, substrate spectrum and kinetics. In complex synthetic systems involving several enzymes, the interrelationship between the affinity and inhibition constants determines the yield and productivity of the syntheses. A detailed analysis and a coordination of enzymes with regard to the above-mentioned approaches has not yet been carried out and is expected to be very complex.

Table 4 summarizes the enzymes which have been used for the synthesis and in situ regeneration of primary and secondary nucleotide sugars as well as recombinant enzyme sources which are available but have not yet been utilized. Enzymes from natural sources were not completely incorporated in Table 4 since they are already reviewed in other publications [189, 195, 260, 261].

In this context, the aim of our work is to provide well defined enzyme systems, to utilize them for the synthesis of these partly instable and very expensive nucleotide sugars, and finally to combine them with glycosyltransferases for glycoconjugate synthesis. In the following sections, I will present a summary of our work on the basis of the biosynthesis pathways for primary and secondary nucleotide sugars.

Fig. 12. Pathways of nucleotide deoxysugars; GDP- and dTDP-activated deoxysugars; ⇨ hypothetical or suggested pathways; ➡ elucidated pathways; E: enzymatically investigated

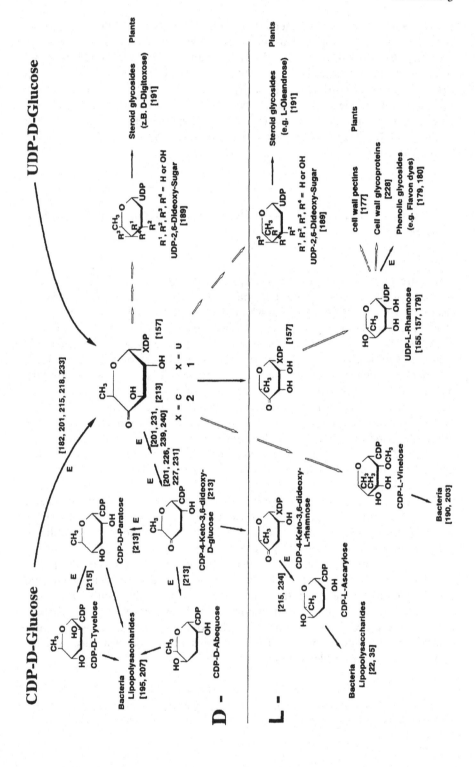

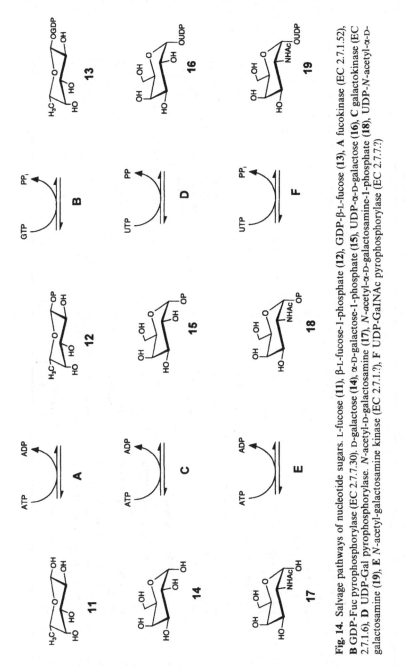

Fig. 14. Salvage pathways of nucleotide sugars. L-fucose (**11**), β-L-fucose-1-phosphate (**12**), GDP-β-L-fucose (**13**), **A** fucokinase (EC 2.7.1.52), **B** GDP-Fuc pyrophosphorylase (EC 2.7.7.30). D-galactose (**14**), α-D-galactose-1-phosphate (**15**), UDP-α-D-galactose (**16**), **C** galactokinase (EC 2.7.1.6), **D** UDP-Gal pyrophosphorylase. N-acetyl-D-galactosamine (**17**), N-acetyl-α-D-galactosamine-1-phosphate (**18**), UDP-N-acetyl-α-D-galactosamine (**19**), **E** N-acetyl-galactosamine kinase (EC 2.7.1.?), **F** UDP-GalNAc pyrophosphorylase (EC 2.7.7.?)

Fig. 13. Pathways of nucleotide deoxysugars; CDP- and UDP-activated deoxysugars; ⇨ hypothetical or suggested pathways; ➡ elucidated pathways; E: enzymatically investigated

Table 4. Purified, partially purified and recombinant enzymes for the synthesis of primary and secondary nucleotide sugars

Nucleotide sugar	Enzyme/EC number	Enzyme source	Nat.	Rec.	In situ Reg.	Synthesized structure	References
Primary Nucleotide Sugars							
UDP-Glc	UDP-Glc Pyrophosphorylase EC 2.7.7.9	yeast,	×		×	Fluorinated Sucrose	[262]
UDP-Glc	UDP-Glc Pyrophosphorylase EC 2.7.7.9	bovine liver	×		×	Nucleotide sugar	[263, 264]
UDP-GlcNH$_2$	UDP-Glc Pyrophosphorylase EC 2.7.7.9	bovine liver, yeast	×			5-Azido-UDP-Glc	[265]
UDP-Glc	UDP-Glc Pyrophosphorylase EC 2.7.7.9	bovine liver		×		—	[266]
UDP-Glc	UDP-Glc Pyrophosphorylase EC 2.7.7.9	potato		×		—	[267]
UDP-Glc	UDP-Glc Pyrophosphorylase EC 2.7.7.9	E. coli K12		×		—	[268]
UDP-Glc	UDP-Glc Pyrophosphorylase EC 2.7.7.9	barley malt	×				[269, 270]
UDP-Glc	UDP-Glc Pyrophosphorylase EC 2.7.7.9	barley malt	×		×	LacNAc	[271]
UDP-Glc	Sucrose Synthase EC 2.4.1.13	rice grains	×			Nucleotide sugar	[272, 273, 274]
UDP-Glc		rice grains	×		×	LacNAc	[273, 275, 276]
UDP-GlcNAc	UDP-GlcNAc Pyrophosphorylase EC 2.7.7.23	calf liver	×			Nucleotide sugar	[277]
UDP-GlcNAc	UDP-GlcNAc Pyrophosphorylase EC 2.7.7.23	E. coli		×		—	[278]
UDP-GlcNAc	UDP-GlcNAc Pyrophosphorylase EC 2.7.7.23	yeast, Candida utilis	×			Nucleotide sugar	[279] [263]
UDP-GlcNAc	UDP-GlcNAc Pyrophosphorylase EC 2.7.7.23	yeast	×		×		[280]
UDP-GlcNAc	UDP-GlcNAc Pyrophosphorylase EC 2.7.7.23	E. coli		×	×	Hyaluronic acid	[281, 282]
dTDP-Glc	dTDP-Glc Pyrophosphorylase EC 2.7.7.24	Salmonella enterica LT2		×	×	—	[283]

Nucleotide sugar	Enzyme / EC number	Source				Product	References
dTDP-Glc	Sucrose Synthase EC 2.4.1.13	rice grains	×			Nucleotide sugar	[273, 274, 284]
CDP-Glc	CDP-Glc Pyrophosphorylase EC 2.7.7.33	Salmonella enterica LT2		×		—	[285]
CDP-Glc	Sucrose Synthase EC 2.4.1.13	rice grains	×			Nucleotide sugar	[272, 273]
ADP-Glc	ADP-Glc Pyrophosphorylase EC 2.7.7.27	potato / E. coli		×		Nucleotide sugar	[286] [287]
ADP-Glc	Sucrose Synthase EC 2.4.1.13	rice grains	×			Nucleotide sugar	[272, 273]
CMP-NeuAc	CMP-NeuAc Synthetase EC 2.7.7.43	E. coli K235	×	×	×	6'-Sialyl-LacNAc	[288, 289]
CMP-NeuAc	CMP-NeuAc Synthetase EC 2.7.7.43	E. coli K235		×	×	3'-Sialyl-LacNAc	[290]
CMP-NeuAc	CMP-NeuAc Synthetase EC 2.7.7.43	calf brain / bovine submaxillary gland	×			Nucleotide sugar	[291, 292] [293, 294, 295]
CMP-NeuAc	CMP-NeuAc Synthetase EC 2.7.7.43	E. coli K235	×			Nucleotide sugar	[296]
CMP-NeuAc	CMP-NeuAc Synthetase EC 2.7.7.43	E. coli O7 K1 / E. coli K235	×	×		Nucleotide sugar	[297] [298]
CMP-N-glycolyl NeuAc CMP-3-deoxy-D-glycero-D-galacto-nonulosonic acid	CMP-NeuAc Synthetase EC 2.7.7.43	calf brain	×			Nucleotide sugar	[299]
CMP-NeuAc derivatives	CMP-NeuAc Synthetase EC 2.7.7.43	calf brain	×			Nucleotide sugar	[291, 299]
CMP-KDO	CMP-KDO Synthetase	calf brain	×	×		Nucleotide sugar	[300, 301] [302]
GDP-Man	GDP-Man Pyrophosphorylase EC 2.7.7.13	yeast / porcine liver	× ×			Nucleotide sugar and derivatives 8-Azido-GDP-Man	[303–308]
GDP-Man	GDP-Man Pyrophosphorylase EC 2.7.7.13		×			GDP-Man	[309]

Table 4. (*Continued*)

Nucleotide sugar	Enzyme/EC number	Enzyme source	Nat.	Rec.	In situ Reg.	Synthesized structure	References
Primary Nucleotide Sugars							
GDP-Man	GDP-Man Pyrophosphorylase EC 2.7.7.13	Pseudomonas aeruginosa		×		—	[310]
GDP-Man	GDP-Man Pyrophosphorylase EC 2.7.7.13	Salmonella enterica LT2		×		Nucleotide sugar	[311]
GDP-Fuc	GDP-Fuc Pyrophosphorylase EC 2.7.7.30	porcine thyroids glands	×			Nucleotide sugar	[312]
GDP-Fuc	GDP-Fuc Pyrophosphorylase EC 2.7.7.30	porcine thyroids glands	×		×	Sialyl-Lewis x	[290]
UDP-Gal, UDP-2-Deoxygal	UDP-Glc 4-Epimerase EC 5.1.3.2	yeast	×			LacNAc and derivatives	[130, 256, 259, 290, 294, 313, 314]
UDP-Gal	UDP-Glc 4-Epimerase EC 5.1.3.2	E. coli		×	×	—	[315]
UDP-Gal	UDP-Glc 4-Epimerase EC 5.1.3.2	Arabidopsis thaliana		×		—	[316]
UDP-Gal, UDP-2-Deoxygal UDP-GalNH₂	Galactose-1-P Uridyltransferase EC 2.7.7.12	yeast	×		×	LacNAc and derivatives	[290, 317]
UDP-Gal, UDP-GalNH₂	Galactose-1-P Uridyltransferase EC 2.7.7.12	yeast	×			Nucleotide sugar	[318, 319]
UDP-Gal (Furanose)	UDP-Galactopyranose Mutase EC 5.4.99.9	E. coli K12		×		Nucleotide sugar	[320]
Secondary Nucleotide Sugars							
UDP-GlcA	UDP-Glc Dehydrogenase EC 1.1.1.22	bovine liver	×		×	Glucuronides	[321]

	Enzyme	Source				References
UDP-GlcA	UDP-Glc Dehydrogenase EC 1.1.1.22	bovine liver	×		5-Azido-UDP-GlcA	[322]
UDP-GlcA	UDP-Glc Dehydrogenase EC 1.1.1.22	bovine liver	×		Nucleotide sugar	[323]
UDP-GlcA	UDP-Glc Dehydrogenase EC 1.1.1.22	E. coli K5 radish	×	×	Hyaluronic acid	[281, 282]
UDP-GalA	UDP-GlcA 4-Epimerase EC 5.1.3.6			×	Nucleotide sugar	[324]
UDP-Xylose	UDP-GlcA Carboxylase EC 4.1.1.35	Wheat germ	×		5-Azido-UDP-Xylose	[325]
UDP-NAc-Muraminic acid pentapeptide	UDP-N-Acetylmuramate- L-Alanin Ligase EC 6.3.2.8	E. coli		×	Nucleotide sugar peptide	[326]
UDP-NAc-Muraminic acid	UDP-N-Acetylglucosamine Enolpyruvyl Transferase EC?	E. coli		×	Nucleotide sugar	[327, 328]
UDP-NAc-Muraminic acid	UDP-N-Acetylenolpyruvoyl-glucosamine Reductase EC 1.1.1.158	E. coli		×	Nucleotide sugar	[329, 330]
GDP-Fuc	"GDP-Fuc Synthesis Enzyme" EC?	Klebsiella pneumonia	×		Sialyl-Lewis x	[290]
dTDP-6-Deoxy-4-ketoglucose	dTDP-Glc 4,6 Dehydratase EC 4.2.1.46	Salmonella enterica LT2		×	Nucleotide sugar	[212, 331, 332]
dUDP-6-Deoxy-4-ketoglucose	dTDP-Glc 4,6 Dehydratase EC 4.2.1.46	Salmonella enterica LT2		×	Nucleotide sugar	[272, 275]
dTDP-L-Rha	dTDP-Glc 4,6 Dehydratase EC 4.2.1.46, dTDP-4-keto-L-Rha 3,5 epimerase EC 5.1.3.13, dTDP-4-keto-L-Rha reductase EC 1.1.1.133	Salmonella enterica LT2		×	Nucleotide Sugar	[212]

Table 4. (*Continued*)

Nucleotide sugar	Enzyme/EC number	Enzyme source	Nat.	Rec.	In situ Reg.	Synthesized structure	References
Secondary Nucleotide Sugars							
CDP-6-Deoxy-4-ketoglucose	CDP-Glc 4,6 Dehydratase EC 4.2.1.45	*Salmonella enterica* LT2		×		Nucleotide Sugar	[333]
CDP-D-Abequose	CDP-Glc 4,6 Dehydratase EC 4.2.1.45, CDP-6-Deoxy-4-ketoglucose 3-Dehydrase EC?, CDP-6-Deoxy-Δ3,4-glucoseen Reductase EC?, CDP-3,6-Dideoxy-4-ketoglucose-4-Reductase EC 1.17.1.1	*Salmonella enterica* LT2		×		Nucleotide Sugar	[333]
CDP-L-Ascarylose	CDP-Glc 4,6 Dehydratase EC 4.2.1.45, CDP-6-Deoxy-4-ketoglucose 3-Dehydrase EC?, CDP-6-Deoxy-Δ3,4-glucoseen 3-Reductase EC?, CDP-3,6-Dideoxy-4-ketoglucose 5-Epimerase EC? CDP-3,6-Dideoxy-L-glycero-D-glycero-4-hexulose-4-Reductase EC?	*Yersinia pseudotuberculosis*		×		—	[19]

Nat. native enzyme; rec. recombinant enzyme; *In situ* Reg. in situ regeneration of nucleotide sugar.

5.1 Primary Nucleotide Sugars

5.1.1 Pyrophosphorylases

Nucleotide sugar synthesis on a gram scale has been accomplished for UDP-$GlcNH_2$ [263] and CMP-NeuAc [292, 298]. UDP-$GlcNH_2$ serves as a precursor for UDP-GlcNAc which was synthesized by chemical acetylation [263]. However, a recombinant UDP-GlcNAc pyrophosphorylase (Table 4) is now available [278, 282] which has been used for in situ UDP-GlcNAc regeneration [281]. Other pyrophosphorylases from bacteria, plants and bovine liver for the UDP-, CDP-, dTDP- and ADP-activation of D-glucose have already been cloned and expressed (Table 4), but not yet used and characterised for nucleotide sugar synthesis.

A comparison of published kinetic data of UDP-glucose pyrophosphorylase (UDP-Glc PP) generally reveal that the enzyme for plant sources has a higher substrate affinity and is less inhibited by substrates and products [334, 335]. We have for the first time isolated UDP-glucose pyrophosphorylase from germinated barley (malt) [269]. Barley malt, which is produced on an industrial scale, is an excellent enzyme source with enzyme activities of 1.4 U/mg protein and 8 U/g malt. In five steps 1158 U UDP-Glc PP was obtained with a yield of 47% and a purification factor of 131. The enzyme preparation is suitable for nucleotide sugar synthesis since contaminating enzymes such as UDP-Glc dehydrogenase and nucleoside-triphosphate-hydrolysing enzymes are not present. Further purification by gel filtration yielded a homogeneous enzyme which was identified as a monomeric protein by SDS-PAGE. The pH optimum of UDP-Glc PP from barley malt is between pH 6.5 and 7.5 and the enzyme shows no loss of activity over 1 to 2 days at different temperatures when stabilized by $0.1 \, mg \, ml^{-1}$ BSA.

The evaluation of UDP-Glc PP from barley malt has been performed by a complete kinetic characterisation, the determination of the substrate specificity and inhibition studies [270]. The kinetic analysis in synthesis direction gave K_m and V_{max} values for UTP and glucose-1-P of 93 µM and 255 $U \, mg^{-1}$ and 74 µM and 228 $U \, mg^{-1}$. UTP gives a substrate surplus inhibition with an inhibition constant of K_{is} of 7.09 mM. In pyrophosphorylsis direction K_m and V_{max} values were determined for UDP-Glc (0.191 mM and 350 $U \, mg^{-1}$) and for inorganic pyrophosphate (PP_i, 0.172 mM and 345 $U \, mg^{-1}$). Inhibition studies revealed that UDP-Glc is a competitive inhibitor (K_i 0.117 mM) for UTP and a non-competitive inhibitor (K_i 0.015 mM) for glucose-1-P. In the pyrophosphorylysis direction UTP is a competitive inhibitor (K_i 0.169 mM) for UDP-Glc; PP_i (0.213 and 0.952 mM) and inorganic phosphate (12.2 and 10.9 mM) is a non-competitive inhibitor of glucose-1-P and UTP. The analysis of these inhibition studies reveals a sequential ordered Bi-Bi enzyme mechanism (Fig. 15). For the enzymatic synthesis of UDP-Glc, it is important that an optimal ratio of Mg^{2+}/UTP between 5 and 10 is maintained. Testing the substrate specificity of UDP-Glc PP by a newly developed assay [336] gave

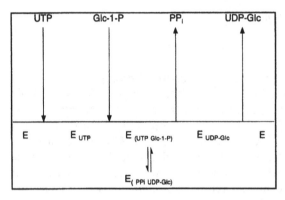

Fig. 15. Sequential ordered Bi-Bi enzyme mechanism of UDP-Glc pyrophosphorylase (EC 2.7.7.9) from barley malt [270]

relative activities for CTP of 18.5%, for GTP of 14.3% and for ATP of 13.7%. The kinetic data of UDP-Glc PP from barley malt confirm that plant enzymes have higher substrate affinities compared to microbial or animal enzymes. The enzyme also shows lower product inhibition than do animal enzymes. The sequentially ordered Bi-Bi enzyme mechanism is a typical feature for the nucleoside diphosphate glucose family [260] to which, besides enzymes from plants, those from animals and yeast also belong [261, 337]. Bacterial enzymes are obviously an exception. The enzyme mechanisms of dTDP-Glc PP and CDP-Glc PP from *Salmonella enterica*, group B, were determined as a ping-pong Bi-Bi mechanism [283, 285]. In combination with whole recombinant yeast cells containing a soluble form of human β1-4 galactosyltransferase, UDP-Glc PP from barley malt was used for the in situ regeneration of UDP-Glc during LacNAc synthesis [271].

GDP-α-D-mannose (**23**) is the donor substrate for mannosyltransferases [139, 146, 338–340] and the precursor of GDP-β-L-fucose (**13**) [173, 197, 243, 341]. Based on the work of Munch-Petersen [342, 343], only crude extracts from yeast have been used for the enzymatic synthesis of labeled and unlabeled **23** and GDP-deoxymannose derivatives (Table 4) [303–305, 307, 308, 344–346] as well as for the in situ regeneration of **23** (Table 4). Common to all these approaches is the use of chemically synthesized sugar-1-phosphates as substrates for GDP-Man PP. An obvious disadvantage of using crude yeast enzyme preparations is the poor quality of the enzyme source since only fresh cells or certain batches of baker's yeast are suitable for synthesis [304, 307]. GDP-Man PP was purified from pig liver and used for the synthesis of 8-Azido-GDP-Man; however, the enzyme lacks absolute specificity for GDP-Man in the pyrophosphorylysis reaction [309].

We have cloned and overexpressed the enzymes phosphomannomutase (PMM, EC 5.4.2.8) and GDP-α-D-mannose pyrophosphorylase (EC 2.7.7.13) from *Salmonella enterica*, group B, in *E. coli* [311]. The partially purified

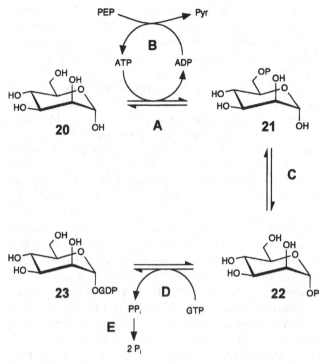

Fig. 16. Enzymatic synthesis of GDP-α-D-mannose (**23**) from D-mannose (**20**) via α-D-mannose-6-phosphate (**21**) and α-D-mannose-1-phosphate (**22**) with in situ regeneration of ATP. **A** Hexokinase (EC 2.7.1.1), **B** Pyruvate kinase (EC 2.7.1.40), **C** Phosphomannomutase (EC 5.4.2.8), **D** GDP-Man pyrophosphorylase (EC 2.7.7.13), **E** inorganic pyrophosphatase (EC 3.6.1.1) [311]

enzymes were characterized and utilized for the synthesis of GDP-Man starting from D-mannose (**20**) (Fig. 16). The genes rfbK (PMM) and rfbM (GDP-Man PP) are located within the rfb cluster coding for the O-antigen biosynthesis. With appropriate primers, both genes were amplified by PCR, cloned and ligated into the pT7-6 expression vector which was induced by IPTG in an *E. coli* BL21 (DE3) or *E. coli* BL21 (DE3) pLysS strain. The specific activity after induction was 0.1 U mg^{-1} for phosphomannomutase and 0.3–0.6 U mg^{-1} for GDP-Man PP. The partially purified GDP-Man PP was proteinchemically and kinetically characterized. The homodimer enzyme (molecular mass 108 kDa) shows a substrate surplus inhibition for GTP (K$_{is}$ 10.9 mM) and mannose-1-P (K$_{is}$ 0.7 mM). GDP-Man is a competitive inhibitor for GTP (K$_i$ 14.7 µM) and an uncompetitive inhibitor for mannose-1-P (K$_i$ 115 µM). This well defined enzyme system was subsequently utilized for enzymatic synthesis in the "repetitive batch" mode. All enzymes were reused in three subsequent batches, and with 80 U of PMM and GDP-man PP, 581 mg (960 µMol, 80% yield) of GDP-man were synthesized. The synthesis was followed by the monitoring of GMP, GDP, GTP and GDP-Man using capillary zone electrophoresis (CZE). The space-time-yield was $2.4 \, gl^{-1} \, d^{-1}$. The protein free solution produced was chromatographed on a Dowex 1×2 (Cl^- form), partially desalted on Sephadex G-10 and lyophilized. The lyophilisate was dissolved in water and **23** was precipitated by the addition of ice-cold acetone. The overall yield was 22.9% (329 µMol, 199 mg) according to CZE.

A de novo synthesis of GDP-Fuc (**13**), UDP-Gal (**16**) and UDP-GalNAc (**19**) by isolated enzymes of the salvage pathways (see Fig. 14) has not yet been described. A crude extract from porcine submaxillary glands gave a yield of 80% for the synthesis of β-L-fucose-1-phosphate (**12**), but the synthesis of **13** was less successful with a yield of 20% [312]. The main pathways for **16** and **19** involve 4-epimerases (Table 4 and Fig. 14) which are not suitable for preparative synthesis due to their unfavourable equilibrium constants [347, 348]. An alternative is Gal-1-P uridyltransferase which is only specific for Gal-1-P and $GalNH_2$-1-P (Table 4).

We have recently established a new continuous microtiter plate assay for screening nucleotide sugar-synthesizing nucleotidyltransferases [336] which also allow us to focus our work on the screening of enzymes from salvage pathways. The assay detects the common product PP_i of all pyrophosphorylases by a PP_i-dependent phosphofructokinase (PP_i-PFK). A subsequent enzyme cascade leads to the production of 2 mol NAD per mol PP_i (Fig. 17). The nucleotidyltransferase substrate screening assay (NUSSA) allows a quick, simultaneous, and comprehensive check of different sugar-1-P and nucleoside triphosphate substrates to be made, using purified pyrophosphorylases or crude extracts from plants, microorganisms, and mammalian tissues [336]. We think that with this assay new enzymes for nucleotide sugar and glycoconjugate synthesis are to be expected.

5.1.2 Sucrose Synthase

We have established a new aspect in nucleotide sugar synthesis by the application of sucrose synthase (UDPG: D-fructose 2 glucosyltransferase, EC 2.4.1.13): the use of sucrose as a renewable resource for nucleotide sugar synthesis [273].

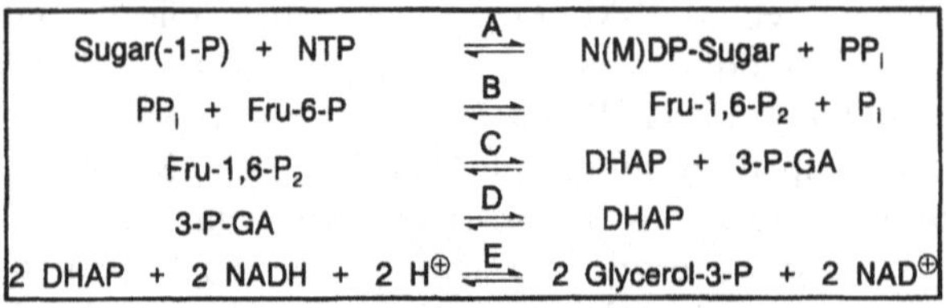

$$\text{Sugar(-1-P)} + \text{NTP} \xrightarrow{\text{A}} \text{N(M)DP-Sugar} + \text{PP}_i$$
$$\text{PP}_i + \text{Fru-6-P} \xrightarrow{\text{B}} \text{Fru-1,6-P}_2 + \text{P}_i$$
$$\text{Fru-1,6-P}_2 \xrightarrow{\text{C}} \text{DHAP} + \text{3-P-GA}$$
$$\text{3-P-GA} \xrightarrow{\text{D}} \text{DHAP}$$
$$\text{2 DHAP} + \text{2 NADH} + \text{2 H}^\oplus \xrightarrow{\text{E}} \text{2 Glycerol-3-P} + \text{2 NAD}^\oplus$$

Fig. 17. Nucleotidyltransferase substrate screening assay (NUSSA). **A** Nucleotide sugar-synthesizing nucleotidyltransferase (EC 2.7.7), **B** PP_i-dependent phosphofructokinase (EC 2.7.2.90), **C** Aldolase (EC 4.1.2.13), **D** Triose-phosphate isomerase (EC 5.3.1.1), **E** Glycerol-3-phosphate dehydrogenase (EC 1.1.1.8) [336]

Sucrose synthase (SuSy) represents a unique plant Leloir glycosyltransferase as it catalyzes a freely reversible reaction in vivo and in vitro (Eq. 1) [349].

$$\text{NDP-Glc} + \text{D-Fructose} \rightleftarrows \text{NDP} + \text{Sucrose} \tag{1}$$

SuSy is one of the major soluble proteins in starch-containing plant organs and plays an important role in the sucrose-starch transformation. The physiological function of SuSy is the cleavage of sucrose with UDP to yield UDP-Glc and fructose [350]. UDP-Glc serves as a precursor in the biosynthesis of starch and as a precursor for secondary nucleotide sugars, e.g. UDP-GlcA, UDP-Ara and UDP-Xyl. In this way SuSy utilizes the energy of the glycosyl bond in sucrose for the synthesis of activated glucoses. The free energy for hydrolysis of sucrose is $\Delta G^0 = -29.3 \text{ KJ mol}^{-1}$ and is close to that of the γ-phosphoryl group of ATP [351]. Therefore, glucose does not need to be activated and isomerised to glucose-6-P and glucose-1-P by hexokinase and phosphoglucomutase, as it is necessary for the activation by pyrophosphorylases (see Fig. 11). In vitro, the pH value can direct the cleavage (pH 6–7) and the synthesis reaction (pH 7–8) of SuSy [350].

Although SuSy has been partially purified and well investigated from many plant sources [352], it has not been recognized as a valuable biocatalyst for the synthesis of nucleotide sugar and saccharides. We have purified sucrose synthase from rice grains [353, 354]. The chosen strategy for the purification of SuSy aimed at the elimination or at least minimization of contaminating enzyme activities such as nucleoside mono- and diphosphatases, invertase and nucleotide sugar hydrolyzing activities in order to make SuSy suitable for nucleotide sugar synthesis. On a pilot scale we obtained 566 U in five steps from 9 kg of rice grains with 11.3% yield and a purification factor of 192 [354]. A crucial point for the proper performance of the chromatography steps was the clarification of the rice homogenate from insoluble starch by pressure filtration. The final sucrose synthase preparation was free of nucleoside mono- and diphosphatases and contained 0.05% of both invertase and UDP-Glc hydrolyzing activities; it could be efficiently utilized for the synthesis of activated glucose. SuSy from rice grains shows a broad substrate spectrum in the cleavage reaction of sucrose (24) with nucleoside diphosphates: UDP > dUDP = dTDP > ADP > CDP > GDP (Fig. 18) [272, 273]. With UDP the sucrose derivative 2-deoxysucrose was cleaved with 55% relative activity [273]. Thus, with one enzyme at least six different activated glucoses, (25–30) can be generated, which is different from the pyrophosphorylases, being quite specific for their nucleoside triphosphate substrates.

The protein-chemical characterisation revealed a homotetrameric enzyme with a molecular mass of 362 kDa, an isoelectric point of pI 6.16 and a blocked N-terminus in Edman degradation studies [353]. In biochemical studies we investigated the effect of free and chelated first-row transition metal ions (Cu^{2+}, Ni^{2+}, Zn^{2+} and Co^{2+}) on SuSy activity [355]. Further experiments on the binding behaviour of SuSy in immobilized metal ion affinity chromatography (IMAC) gave an insight into the topography of sucrose synthase from rice

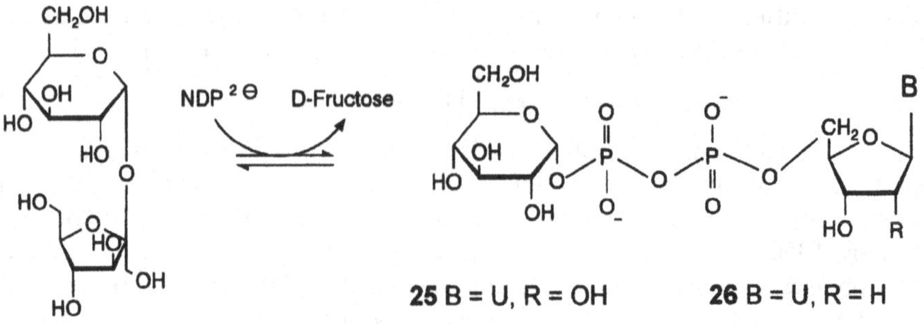

Fig. 18. Substrate spectrum of sucrose synthase (EC 2.4.1.13) from rice grains in the cleavage reaction of sucrose (**24**) with nucleoside diphosphates [272, 273]

and revealed that at least 5–7 surface accessible histidines are present for interaction with immobilized metal ions [355]. These results have an important impact on the application of IMAC for the homogeneous purification of sucrose synthase, as well as on the evaluation of cloning and expression strategies using poly-histidine tails.

dTDP-Glc (**27**) as the precursor of many dTDP-activated deoxysugars (see Fig. 12) was our first target for the preparative synthesis of nucleotide sugars utilizing sucrose synthase. We determined the reaction conditions with respect to the choice of buffer, the pH value, the reaction temperature and the enzyme kinetics [274]. The most important result was that SuSy is inhibited at high concentrations of UDP or dTDP [274]. Taking this result into account we performed the continuous synthesis of **27** in an enzyme-membrane reactor (EMR) to obtain high enzyme productivities [284]. The comparison of the EMR synthesis with an optimized batch synthesis revealed that in the EMR a higher enzyme productivity could be reached by a higher yield in a shorter residence time (Fig. 19). The obtained space-time-yield of $98.1 \text{ g l}^{-1} \text{d}^{-1}$ and the halflife time of 119 h for SuSy means that SuSy can be exploited for gram scale synthesis (ca. 1 g d^{-1} in a 10 ml EMR) of this important nucleotide sugar.

However, nucleoside diphosphates (NDP) are still expensive substrates, which can be obtained from much more cheaper nucleoside monophosphates (NMP). In this respect we have combined the SuSy-catalyzed cleavage of sucrose with the enzymatic formation of NDPs from NMPs catalyzed by nucleoside monophosphate kinase (NMPK, EC 2.7.4.4) or myokinase (MK, EC 2.7.4.3), including in situ regeneration of ATP with pyruvate kinase (PK, EC 2.7.1.40) (Fig. 20) [272]. Testing the substrate spectrum of four different kinases disclosed that none of them accepted dTMP as substrate [272]. However, dUMP was well accepted by NMPK and dUDP-activated glucose could also substitute dTDP-activated glucose as precursor for the synthesis of activated deoxysugars (see below). The excellent enzyme stabilities under synthesis

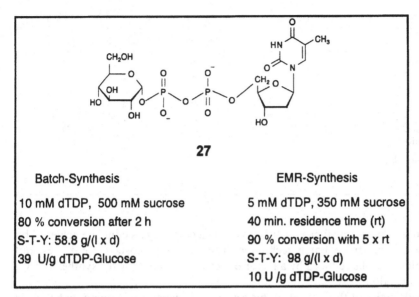

27

Batch-Synthesis	EMR-Synthesis
10 mM dTDP, 500 mM sucrose	5 mM dTDP, 350 mM sucrose
80 % conversion after 2 h	40 min. residence time (rt)
S-T-Y: 58.8 g/(l x d)	90 % conversion with 5 x rt
39 U/g dTDP-Glucose	S-T-Y: 98 g/(l x d)
	10 U /g dTDP-Glucose

Fig. 19. Enzymatic synthesis of dTDP-Glc (**27**) with sucrose synthase in an optimized batch and an enzyme membrane reactor [284]

25 B = U, R = OH   **26** B = U, R = H

28 B = A, R = OH   **29** B = C, R = OH

Fig. 20. Enzymatic synthesis of activated glucoses (**25, 26, 28, 29**) starting from nucleoside monophosphates and sucrose (**24**). **A** nucleoside monophosphate kinase (EC 2.7.7.4) or myokinase (EC 2.7.4.3), **B** sucrose synthase (EC 2.4.1.13)

conditions in the presence of 1 mg ml^{-1} BSA and 500 mM sucrose facilitated the synthesis of ADP-, CDP-, UDP- and dUDP-Glc with the repetitive batch technique in order to obtain maximum enzyme productivities. The enzymes were recovered by ultrafiltration after each batch and repetitively used by the addition of a fresh substrate solution. Table 5 summarizes the synthesis of activated glucoses starting from sucrose and NMP. Isolation of the product was achieved by chromatography of the alkaline-phosphatase-treated and ultrafiltrated product solution on Sepharose Q FF (Cl$^-$ form) or Dowex 1 x 2 (100–200 mesh, Cl$^-$ form) and desalting on Sephadex G-10. The structural integrity of the synthesized nucleotide sugars were confirmed by ^{1}H- and ^{13}C-NMR.

Table 5. Enzymatic synthesis of activated glucoses starting from sucrose und NMP [272]

	28	29	25	26
Batches [–]	3	5	6	8
Time [h]	33	107	133	64
Yields [%]				
Synthesis	39	7	38	68
Isolation	27	19	57	72
Overall	11	1, 3	22	49
Product [mg]	76, 7	9, 2	141	162
Enzymes [U]				
SuSy	100	50	6.3	40
NMPK/Mk	-/50	-/500	5/-	20/-

In summary, an efficient synthesis of nucleotide sugars could be achieved by using the repetitive batch technique. The estimated costs for enzymes and chemicals can be reduced by a factor of between 4 and 8 for the synthesis of 1 g of activated glucose [272].

Although not the subject of this review, it is worth mentioning that SuSy is also a valuable catalyst for the synthesis of sucrose analogs [273, 356, 357].

6 Enzymatic Synthesis of Secondary Nucleotide Sugars

Table 4 presents a summary of the enzymes from different nucleotide sugar pathways and reflects the possible current availability of secondary nucleotide sugars for glycoconjugate synthesis with Leloir glycosyltransferases. In the following sections I will only refer to examples where a synthesis of the corresponding secondary nucleotide sugars has been performed, with the exception of in situ regeneration systems (see below).

6.1 UDP-Activated Secondary Nucleotide Sugars

UDP-Gal and UDP-GalNAc as the donor substrates for e.g. core 2 protein O-glycosylation are in vivo generated by enzymatic 4-epimerization of UDP-Glc and UDP-GlcNAc or enzymes of the salvage pathways (see above). Since both approaches are limited by unfavourable equilibrium constants or the availability of the enzymes, a partially chemoenzymatic approach has been developed [318]. The enzyme gal-1-P uridyltransferase (EC 2.7.7.12) provides UDP-Gal or UDP-GalNH$_2$ starting with Gal-1-P and GalNH$_2$-1-P, respectively. UDP-Glc serves as donor for UMP and is regenerated from liberated Glc-1-P by UDP-Glc pyrophosphorylase. In a chemical step, UDP-GalNH$_2$ is acetylated by N-acetoxysuccinimide without prior isolation. The yields of UDP-Gal and UDP-GalNAc, starting from Gal-1-P and GalNH$_2$-1-P, were 43% and 34%, respectively [318]. We have developed a similar chemoenzymatic synthesis of UDP-GalNAc starting from the readily available substrates sucrose (24) and UMP (Fig. 21) [319], thus avoiding the stoichiometric consumption and in situ regeneration of UDP-Glc. In the enzymatic reaction (Fig. 22), we combined NMPK, forming UDP from UMP with sucrose synthase generating UDP-Glc (25) from UDP and sucrose (24) and Gal-1-P uridyltransferase to yield UDP-GalNH$_2$ (33) with GalNH$_2$-1-P as acceptor for UMP. Since the equilibrium constant of Gal-1-P uridyltransferase with the unnatural substrate GalNH$_2$-1-P (31) ($K_{eq} = 0.26$) is unfavourable compared to Gal-1-P ($K_{eq} = 1.1$), the equilibrium was shifted by further conversion of Glc-1-P (32) to gluconate-6-P with in situ regeneration of NAD. NAD regeneration was enzymatically coupled to the regeneration of UTP for the NMPK reaction from PEP by pyruvate kinase (Fig. 22). The yield of the enzymatic reaction was 55.8% with respect to UMP; the overall yield of the chemoenzymatic reaction after product isolation was 45.1% with respect to UMP [319].

UDP-GlcNAc represents the branching point for the bacterial biosynthesis of murein and lipopolysaccharides where secondary nucleotide sugars are involved [329, 358]. Starting from fructose-6-P, all enzymes are available as recombinant proteins and enzymatic synthesis of important intermediates seems to be possible.

Photoaffinity analogs of UDP-Xyl and UDP-GlcA have been chemoenzymatically synthesized for a covalent derivatization and ^{32}P-labeling of membrane-bound glycosyltransferases and nucleotide sugar metabolizing enzymes (Table 4) [322, 325, 359, 360].

6.2 Synthesis of Nucleoside Diphosphate Activated Deoxysugars

Recombinant enzyme systems from the rfb-cluster of enterobacteria are available for the synthesis of the bacterial nucleotide deoxysugars dTDP-L-rhamnose, CDP-D-abequose and CDP-L-ascarylose starting from dTDP-D-Glc or CDP-D-Glc (Table 4). Glycosyltransferases can now be identified for in vitro synthesis of bacterial O-antigen structures. In the case of dTDP-activated deoxysugars, the important intermediate dTDP-6-deoxy-4-ketoglucose (37) has also been synthesized from dTDP-Glc (27) with recombinant dTDP-Glc 4,6 dehydratase (EC 4.2.1.46) on a 0.2 gram scale [212]. The intermediate was obtained with 89% yield and characterized by HPLC, GC and GC-MS. In an additional work, we have synthesized dTDP-6-deoxy-4-ketoglucose (37) with recombinant dTDP-Glc 4,6 dehydratase for characterization by NMR and ESI-MS [331]. We also obtained data concerning the substrate spectrum of the wild-type enzyme from E. coli B, resulting in the preparative synthesis of dTDP-activated 3,6-dideoxy-4-ketoglucose (38) and dTDP-3-azido-3,6-dideoxy-4-ketoglucose (39) (Fig. 23) [331]. With a newly developed analytical

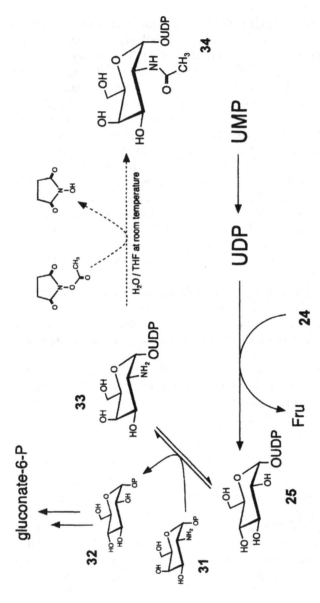

Fig. 21. Chemoenzymatic synthesis of UDP-GalNAc (**34**) starting from UMP and sucrose (**24**) [319]

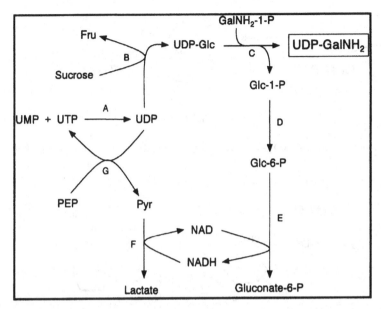

Fig. 22. Schematic presentation of the enzymatic synthesis of UDP-GalNH$_2$ (**33**) including cofactor regeneration systems. **A** nucleoside monophosphate kinase (EC 2.7.7.4), **B** sucrose synthase (EC 2.4.1.13), **C** gal-1-P uridyltransferase (EC 2.7.7.12), **D** phosphoglucomutase (EC 2.7.5.1), **E** glucose-6-P dehydrogenase (EC 1.1.1.49), **F** lactate dehydrogenase (EC 1.1.1.27), **G** pyruvate kinase (EC 2.7.1.40) [319]

and preparative HPLC method, which is based on ion-pair reversed-phase (IP-RP) HPLC, it was possible to separate dTDP-activated hexoses [331]. We found that **37** was unstable when stored in the presence of salt or buffer salts at − 4°C or − 20°C, after lyophilization from volatile or salt-containing buffers as well as after precipitation with organic solvents. **37** was isolated by preparative IP-RP HPLC and the desalted product is stable at − 20°C for at least four months. The spectroscopic characterization revealed that the pure product was obtained as 1:5 mixture of the keto- and hydrate-form [331].

We have also established the enzymatic synthesis of dTDP-6-deoxy-4-ketoglucose on a gram scale starting from dTDP and sucrose by the combination of sucrose synthase with recombinant dTDP-Glc 4,6 dehydratase [332] (Fig. 24). Since dTDP-Glc 4,6 dehydratase is competitively inhibited by dTDP a fed-batch technique was chosen for the large scale synthesis. In five batches, 2.26 mmol of product (96% yield) was synthesized within 5 h with 53 U SuSy and 81 U dTDP-Glc 4,6 dehydratase which corresponds to a space-time-yield of 133 g l^{-1} d^{-1}. After isolation of the product by ion exchange chromatography on Dowex 1 × 2 (Cl$^-$ form), desalting on Sephadex G-10 and lyophilisation, 1033 mg of dTDP-6-deoxy-4-ketoglucose were obtained (overall yield 74.5%) [332].

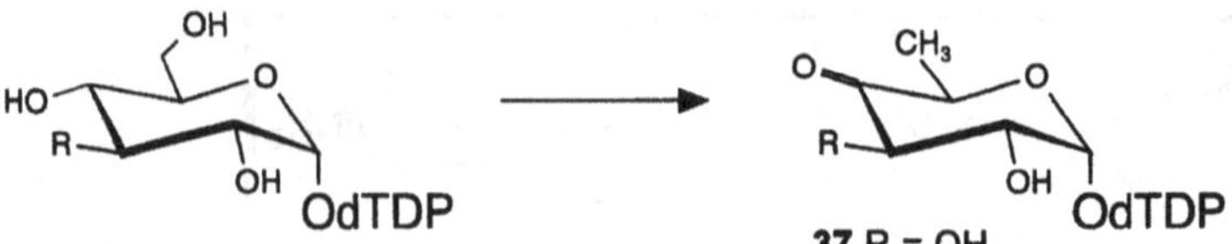

27 R = OH
35 R = H
36 R = N_3

37 R = OH
38 R = H
39 R = N_3

Fig. 23. Enzymatic synthesis of dTDP-activated deoxysugars with dTDP-Glc 4,6 dehydratase (EC 4.2.1.46) [331]

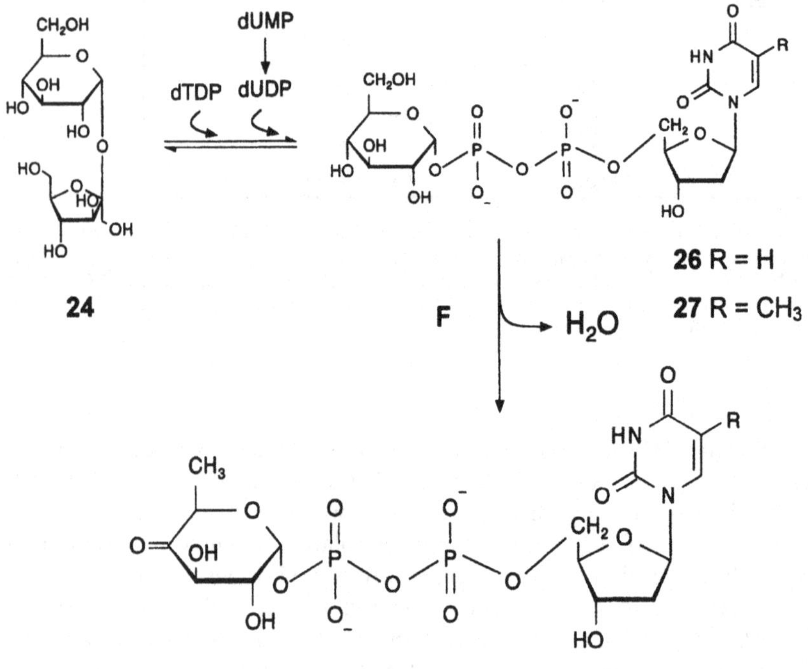

24

26 R = H
27 R = CH_3

40 R = H
37 R = CH_3

Fig. 24. Enzymatic synthesis of dTDP- and dUDP-activated 6-deoxy-4-ketoglucose (37, 40) starting from sucrose (24) and nucleoside di- or monophosphates [272, 332]. F recombinant dTDP-Glc 4,6 dehydratase

We also combined further the enzymatic synthesis of dUDP from dUMP by NMPK with SuSy and dTDP-Glc 4,6 dehydratase (Fig. 24) [272]. Using again the repetitive batch technique for enzymatic synthesis, 120 mg dUDP-6-deoxy-4-ketoglucose was obtained after product isolation (49% overall yield).

Both products are now available as tools for the investigation of deoxysugar pathways (Fig. 12).

7 In Situ Regeneration of Nucleotide Sugars

An alternative to the synthesis of nucleotide sugars on a large scale is the in situ regeneration of the activated sugars. Fig. 25 shows the enzymatic synthesis of N-acetyllactosmine (LacNAc) with in situ regeneration of UDP-Glc and UDP-Gal [259]. In this cycle, the reaction of β1-4 galactosyltransferase (β1-4 GalT) is coupled to the regeneration system for UDP-Glc and UDP-Gal consisting of four enzymes. The nucleoside diphosphate UDP, which is an inhibitor of β1-4 GalT, is converted to UDP-Glc by pyruvate kinase and UDP-Glc pyrophosphorylase. The equilibrium of the pyrophosphorylase reaction is shifted by the cleavage of inorganic pyrophosphate by pyrophosphatase. UDP-Glc 4-epimerase forms UDP-Gal, the donor substrate for β1-4 GalT. The exchange of UDP-Glc 4-epimerase by gal-1-P uridyltransferase (Fig. 26) gives a better equilibrium constant for the formation of UDP-Gal ($K_{eq} = 1.1$) and makes the input of modified galactose derivatives possible [317].

Further in situ regeneration systems were established for UDP-GlcNAc [281], UDP-GlcA (281, 321), GDP-Man [340], GDP-Fuc [290] and CMP-NeuAc [288, 289]. The advantage of regeneration cycles for nucleotide sugars is that, apart from the removal of inhibitory nucleoside diphosphates, only small amounts of nucleotide sugars are generated during the reaction and the complex isolation of activated sugars is not necessary. The cycles are energetically driven by phospho(enol) pyruvate, which is consumed in equimolar amounts by the acceptor substrate of the glycosyltransferase [256]. Table 4 also summarizes the carbohydrate structures which have been synthesized by glycosyltransferases

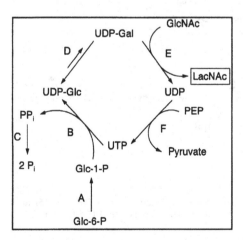

Fig. 25. Enzymatic synthesis of LacNAc with in situ regeneration of UDP-Glc and UDP-Gal. **A** phosphoglucomutase (EC 2.7.5.1), **B** UDP-Glc pyrophosphorylase (EC 2.7.7.9), **C** inorganic pyrophosphatase (EC 3.6.1.1), **D** UDP-Glc 4-epimerase (EC 5.1.3.2), **E** β1-4 galactosyl transferase (EC 2.4.1.38), **F** pyruvate kinase (EC 2.7.1.40) [259]

with in situ regeneration of their donor substrates. Recombinant enzyme systems for in situ regeneration of UDP-GlcNAc, UDP-GlcA, and CMP-NeuAc have been used for the synthesis of hyaluronic acid and sialylated glycoconjugates (Table 4). All other enzyme systems refer to native enzymes, some of which are only partially purified, e.g. GDP-Man and GDP-Fuc generating enzymes stem from *Saccharomyces cerevisiae* and *Klebsiella pneumoniae* or porcine thyroid glands [256].

We have developed a novel enzymatic reaction cycle for the synthesis of LacNAc with in situ regeneration of UDP-Glc (**25**) and UDP-Gal (**16**) (Fig. 27) [273, 275]. The unique character of sucrose synthase to generate **25** by the cleavage of sucrose (**24**) with UDP makes this enzyme more favourable compared with pyrophosphorylases. In this way UDP-Glc is directly regenerated and further epimerized to **16** by UDP-Glc 4-epimerase from yeast, yielding the donor substrate for human milk β1-4 GalT. In comparison to other published

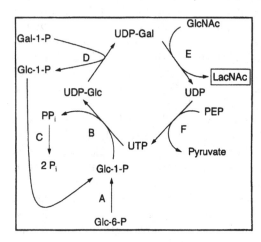

Fig. 26. Enzymatic synthesis of LacNAc with in situ regeneration of UDP-Glc and UDP-Gal. **A** phosphoglucomutase (EC 2.7.5.1), **B** UDP-Glc pyrophosphorylase (EC 2.7.7.9), **C** inorganic pyrophosphatase (EC 3.6.1.1), **D** gal-1-P uridyltransferase (EC 2.7.7.12), **E** β 1-4 galactosyltransferase (EC 2.4.1.38), **F** pyruvate kinase (EC 2.7.1.40) [317]

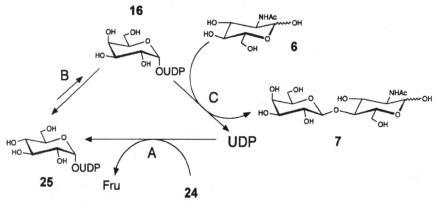

Fig. 27. Enzymatic synthesis of LacNAc (**7**) with in situ regeneration of UDP-Glc (**25**) and UDP-Gal (**16**). **A** sucrose synthase (EC 2.4.1.13), **B** UDP-Glc 4-epimerase (EC 5.1.3.2), **C** β 1-4 galactosyltransferase (EC 2.4.1.38) [273, 275]

cycles (see Figs. 25 and 26), PEP is not necessary for nucleoside triphosphate regeneration nor is inorganic phosphate (P_i) generated excluding enzyme inhibition by P_i. Three objectives have been followed in order to optimize reaction conditions for the novel reaction cycle [275]: (i) high enzyme productivities (g LacNAc per U enzyme; (ii) repetitive use of all enzymes to increase enzyme productivity and decrease costs for the enzymes; (iii) reactivation of reductively inactivated UDP-Glc 4-epimerase, occuring in the presence of UMP and monosaccharides, which are acceptor substrates of β1-4 GalT.

The reaction conditions for an optimum cooperation of all enzymes were determined by sequential and parallel optimization strategies, resulting in a 100% conversion of 10 mM GlcNAc (6) within 15 h [275]. The enzyme productivity of β1-4 GalT, the most expensive enzyme of the cycle, was 0.077 g U^{-1}. To increase enzyme productivity all soluble enzymes were repetitively used by the repetitive batch technique. Although the first batch gave 95% after 12 h as expected, the second batch gave only 6.2% conversion after 8 h. A check on all enzyme activities revealed that the UDP-Glc 4-epimerase was no longer active. The addition of fresh enzyme raised the conversion to 67.1% after 14 h, but this was certainly not the best way to increase enzyme productivity.

A solution for this problem has been deduced from the extensively investigated enzyme mechanism of UDP-Glc 4-epimerase from yeast and E. coli. The enzyme is inactivated by a so-called "suicide" mechanism [347, 361, 362]. The homodimeric enzyme needs NAD^+ as cofactor, which is tightly bound to the enzyme by non-covalent forces. The C-4 position of the nucleotide sugar is epimerized, via UDP-4-ketoglucose as a transition state, with the concomitant reduction of NAD^+ (Fig. 28). With the subsequent reduction of the 4-oxo group, the cofactor is regenerated intramolecularly. UDP-Glc 4-epimerase is reductively inactivated in the presence of UMP and different monosaccharides (e.g. 42) [347, 361, 362]. The binding of UMP at the enzyme's nucleotide binding site induces a conformational change of the protein, increasing the reactivity of the

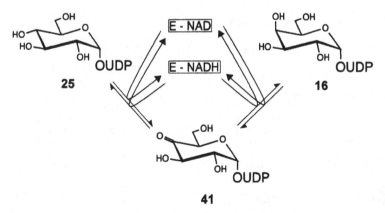

Fig. 28. Reaction mechanism of UDP-Glc 4-epimerase (EC 5.1.3.2). E-NAD enzyme-bound cofactor

cofactor towards reducing monosaccharides. The oxidized monosaccharides (e.g. **43**) leave the enzyme very quickly due to its low binding constant and the enzyme-bound NADH cannot be regenerated (Fig. 29). In a similar way, the substrates UDP-Glc (**25**) or UDP-Gal (**16**) can also lead to reductive enzyme inactivation by displacement of the transition state. The presence of UMP as a by-product in enzymatic reaction cycles with glycosyltransferases is a general problem since nucleotides are chemically degraded by the catalytic action of Mn^{2+} ions, which serve as cofactors of glycosyltransferases [363]. In the presence of UMP and derivatives of glucose, which are acceptor substrates of β1-4, GalT, UDP-Glc 4-epimerase is inactivated (Table 6). In the presence of their substrates (1 mM) UDP-Glc 4-epimerase loses 40% of its activity after 24 h. In summary, UDP-Glc 4-epimerase is reductively inactivated in LacNAc synthesis cycles by the concerted effect of UMP and glucose or derivatives thereof as well as by UDP-Glc or UDP-Gal alone.

Reactivation of UDP-Glc 4-epimerase can be achieved by incubation with the transition state analogs, dUDP- or dTDP-6-deoxy-4-ketoglucose (**40**, **37**), resulting in an enzyme-bound cofactor regeneration (Fig. 30). The full activity of UDP-Glc 4-epimerase was recovered and the stability of the enzyme increased by incubation with these transition state analogs which have been synthesized by us on a 0.1 to 1 g scale [272, 331, 332] (see also Sections 6.1 and 6.2). With this

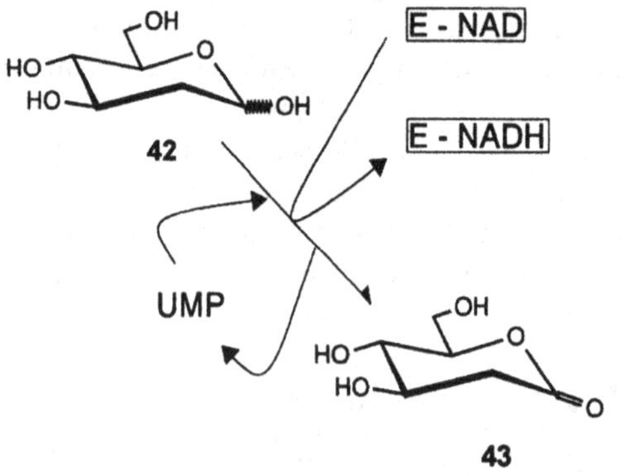

Fig. 29. Reductive inactivation of UDP-Glc 4-epimerase (EC 5.1.3.2)

Table 6. Reductive Inactivation of UDP-Glc 4-epimerase by different monosaccharides and derivatives thereof [275]

Saccharide	Relative activity (%)	Incubation time (h)
N-Octylglucopyranoside	29.3	24
5-Thioglucose	34.3	7
Glucose	15.8	2
2-Deoxyglucose	20.7	2
6-Aminohexyl-GlcNAc	100	7
GlcNAc	97.5	7
UDPGlc/UDP-Gal	58.3	24

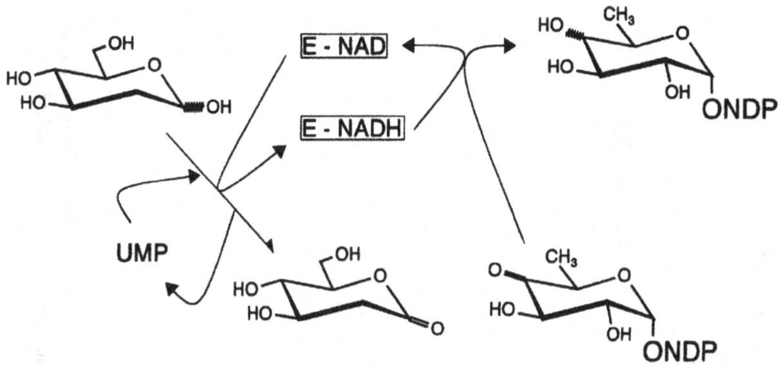

Fig. 30. Reactivation of UDP-Glc 4-epimerase (EC 5.1.3.2) with the transition state analogs dUDP-6-deoxy-4-ketoglucose (N = U, **26**) and dTDP-6-deoxy-4-ketoglucose (N = T, **27**)

Table 7. Comparison of parallel and repetitive batches for the enzymatic synthesis of LacNAc (**7**) [275]

	Parallel batches	Repetitive batches
Batches/repetitions	11/0	1/10
β1-4 GalT (U)	13.75	1.25
UDP-Glc 4-epimerase (U)	55	5
Sucrose synthase (U)	110	10
Average yield (%)	100	57.4
Space-time-yield (g l^{-1} day^{-1})	6.1	2.2
Cycle number	10	5.7
synthesized **7** (g)	1.05	0.594
g **7**/U GalT	0.077	0.475
g **7**/U epimerase	0.019	0.119
g **7**/U SuSy	0.010	0.059

improved stability of UDP-Glc 4-epimerase, LacNAc was synthesized by the novel reaction cycle (see Fig. 27) using the repetitive batch technique (Table 7). In 11 successive batches, 594 mg (1.55 mmol) LacNAc was synthesized using only 5 U UDP-Glc 4-epimerase, 10 U SuSy, 1.25 U β1-4 GalT and 14.3 mg (0.027 mmol) of dUDP-6-deoxy-4-ketoglucose. Although the average yield, the cycle number and the space-time yield were lower compared to parallel batches, the enzyme productivities were increased and the costs for enzymes and chemicals decreased by a factor of 6, respectively (Table 7). LacNAc was isolated in five steps with an overall yield of 34.4% (356 mg). The procedure included enzymatic degradation of sucrose by invertase, separation of the product from D-fructose, Glc, and GlcNAc on a AG50W-X8 column (Ca^{2+} form), removal of buffer salts and other charged substances by anion exchange chromatography on Dowex 1 × 2, 100–200 mesh, Cl$^-$ form), and desalting by a gel filtration step.

The novel three-enzyme-reaction cycle was further combined with recombinant α1-3 galactosyltransferase to synthesize the important trisaccharide

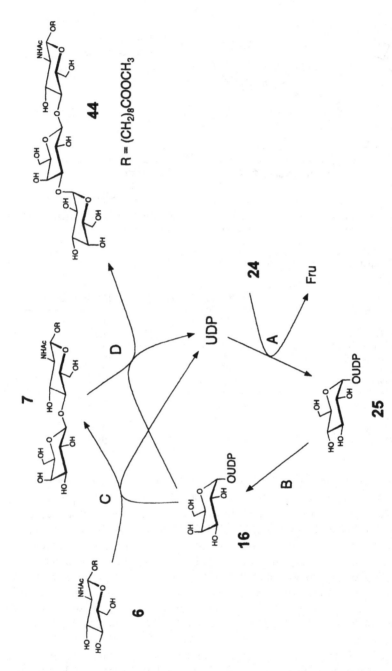

Fig. 31. Enzymatic synthesis of Gal(α1-3) Gal(β1-4) GlcNAc(β1-)O-(CH₂)₈COOHCH₃ (**44**) with in situ regeneration of UDP-Glc (**25**) and UDP-Gal (**16**). **A** sucrose synthase (EC 2.4.1.13), **B** UDP-Glc 4-epimerase (EC 5.1.3.2), **C** β1-4 galactosyltransferase (EC 2.4.1.38), **D** α1-3 galactosyltransferase (EC 2.4.1.124/151)

epitope Gal(α1-3)Gal(β1-4) GlcNAc(β1-)O-$(CH_2)_8COOCH_3$ (**44**) with four enzymes, obtaining a 85% yield (Fig. 31) [276]. The β1-4 GalT reaction purchases the acceptor substrate LacNAc (**7**) for the α1-3 GalT reaction. The in situ regeneration cycle for the donor substrate UDP-Gal (**16**) serves both glycosyltransferases, reacting in successive order. This is the first example where one in situ regeneration cycle for a nucleotide sugar is combined with two glycosyltransferase using the same donor substrate.

In summary, many important in situ regeneration cycles for nucleotide sugars have been established for the synthesis of glycoconjugates with glycosyltransferases. However, more detailed investigations will be needed to define the inhibition, the stability and the kinetics of the enzymes involved, and to achieve synthesis conditions with high yields and high enzyme productivities.

8 Conclusions and Outlook

The application of enzymes for the synthesis of primary and secondary nucleotide sugars is only one part of the complex field of glycobiotechnology. In combination with Leloir-glycosyltransferases they can be recommended as very effective tools for producing important di- or oligosaccharide structures. In this context, the establishment of well-defined and characterized enzyme systems for the synthesis of nucleotide sugars is as important as the availability of glycosyltransferases. In many cases, the availability of nucleotide deoxysugars as donor substrates makes the identification and biochemical characterization of Leloirglycosyltransferases even possible. In future, more of these complex enzyme systems are to be expected because of the important role played by deoxysugars in glycoconjugates.

Acknowledgements. I wish to thank Prof. Dr. Maria-Regina Kula for the excellent support she gave to our work and for her continuing personal interest. I also gratefully acknowledge the considerable efforts of my co-workers, who are listed as co-authors of our publications. I wish to thank Prof. C. Wandrey and Dr. U. Kragl (Institute for Biotechnology 2, Research Center Jülich, Germany), Prof. W. Piepersberg, Dr. J. Distler and Dipl. Biol. S. Verseck (Institute of Chemical Microbiology, University Wuppertal, Germany), Prof. Dr. D.H. Van den Eijnden, Dr. C.H. Hokke and Dr. A.P. Neeleman (Department of Medical Chemistry, Vrije Universiteit Amsterdam, The Netherlands) for their excellent cooperation. Financial support from the Hoechst AG, Frankfurt am Main, Germany, the Deutsche Forschungsgemeinschaft (DFG, project El 135/2-1) and the EU within the 4th Framework Biotechnology 1994–1998 (project: "Engineering protein O-glycosylation for the production of receptor blockers", contract no. BIO4-CT95-0138) is also gratefully acknowledged.

Note added in proof

After preparation of this review, an abstract by Gödde A, Kragl U, Biselli M, Katapodis A, Bowen BR, Ernst B, Wandrey C, was published on the XVIIIth International Carbohydrate Symposium, July 21–26, 1996, Milano, Italy (BP 135, p 390). It reports the large-scale production of 1800 units of a recombinant human α1-3 fucosyltransferase (FucT VI, EC. 2.4.1.152) in its soluble form with a CHO cell line in a continuous fermentation process in a fluidized bed reactor ($34\,\mathrm{mU\,mg}^{-1}$ in culture supernatant). After partial purification the specific activity yielded $220\,\mathrm{mU\,mg}^{-1}$.

9 References

1. Montreuil J (1995) The history of glycoprotein research, a personal view. In: Montreuil J, Schachter H, Vliegenthart JFG (eds) Glycoproteins New Comprehensive Biochemistry Vol 29a. Elsevier, Amsterdam, p 1
2. Feizi T, Larkin M (1990) Glycobiology 1: 17
3. Colman PM (1994) Protein Science 3: 1687
4. Jann K, Jann B (1984) Structure and biosynthesis of O-antigens. In: Rietschel ET (ed) Handbook of endotoxin Vol 1 Chemistry of endotoxin. Elsevier, Amsterdam, p 138
5. Whitfield C, Valvano MA (1993) Adv Microb Physiol 35: 135
6. Zähringer U, Lindner B, Rietschel ET (1994) Adv Carbohydr Chem Biochem 50: 211
7. Tomlinson S, De Carvalho LP, Vandekerckhove F, Nussenzweig, V (1992) Glycobiology 2: 549
8. Vandekerckhove F, Schenkman S, De Carvalho LP, Tomlinson S, Kiso M, Yoshida M, Hasegawa A, Nussenzweig V (1992) Glycobiology 2: 541
9. Van den Eijnden DH, Neeleman AP, Van der Knaap WPW, Bakker H, Agterberg M, Van Die I (1995) Biochem Soc Trans 23: 175
10. Van den Eijnden DH, Neeleman AP, Van der Knaap WPW, Bakker H, Agterberg M, Van Die I (1995) Adv Exp Med Biol 376: 47
11. Paulson JC (1992) Selectin/carbohydrate mediated adhesion of leukocytes. In: Harlan JM, Liu DY (eds) Adhesion: Its role in inflammatory disease. Freeman, New York, p 19
12. Varki A (1994) Proc Natl Acad Sci USA 91: 7390
13. Lasky LA (1995) Ann Rev Biochem 64: 113
14. Furukawa K, Kobata A (1992) Cell surface carbohydrates – their involvement in cell adhesion. In: Ogura H (ed) Carbohydrates – synthetic methods, applications in medicinal chemistry. VCH, Weinheim, p 369
15. Hakomori S-I (1984) Am J Clin. Pathol 82: 635
16. Feizi T (1985) Nature 314: 53
17. Fukuda M (1994) Carbohydrate-mediate adhesion of tumor cells. In: Bock K (ed) Complex carbohydrates in drug research, Alfred Benzon Symposium. Mungsgaard, Copenhagen, p 353
18. Piepersberg W (1994) CRC Critical Reviews in Biotechnology 14: 251
19. Liu H-W, Thorson JS (1994) Ann. Rev Microbiol 48: 223
20. Baenziger JU, Green ED (1988) Biochim Biophys Acta 947: 287
21. Cumming DA (1991) Glycobiology 1: 115
22. Rasmussen JR (1991) Glycosylation of recombinant proteins. In: Ginsberg V, Robbins PW (eds) Biology of Carbohydrates Vol 3, Jai Press, London, p 179
23. Takeuchi M, Kobata A (1991) Glycobiology 1: 337
24. Hart G, Holt GD, Haltiwanger RS (1988) TIBS 13: 380
25. Montreuil J, Vliegenthart JFG, Schachter H (eds) (1995) New Comprehensive Biochemistry, Vol 29a, Glycoproteins. Elsevier, Amsterdam

26. Fukuda M, Hindsgaul O (eds) (1994) Frontiers in Molecular Biology, Molecular Glycobiology. IRL Press, Oxford
27. Van Echten G, Sandhoff K (1993) J Biol Chem 268: 5341
28. Varki A (1993) Glycobiology 3: 97
29. Lis H, Sharon N (1993) Eur J Biochem 218: 1
30. Rademacher TW, Parekh RB, Dwek RA (1988) Ann Rev Biochem 57: 785
31. Kornfeld R, Kornfeld S (1985) Ann Rev Biochem 54: 631
32. Tanner W, Lehle L (1987) Biochim Biophys Acta 906: 81
33. Watkins WM (1991) Pure Appl Chem 63: 561
34. Schmidt RR (1986) Angew Chem Int Ed Engl 25: 212
35. Paulsen H (1982) Angew Chem Int Ed Engl 21: 155
36. Kondo H, Aoki S, Ichikawa Y, Halcomb RL, Ritzen H, Wong C-H (1994) J Org Chem 59: 864–877
37. Schmidt RR, Kinzy W (1994) Adv Carbohydr Chem Biochem 50: 21
38. Garg HG, von dem Bruch K, Kunz H (1994) Adv Carbohydr Chem Biochem 50: 277
39. Khan SH, Hindsgaul O (1994) Chemical synthesis of oligosaccharides. In: Fukuda M, Hindsgaul O (eds) Frontiers in Molecular Biology, Molecular Glycobiology. IRL Press, Oxford, p 206
40. Gijsen HJM, Qiao L, Fitz W, Wong C-H (1996) Chem Rev 96: 443
41. Wong C-H, Halcomb RL, Ichikawa Y, Kajimoto T (1995) Angew Chem Int Ed Engl 34: 412
42. Wong C-H, Halcomb RL, Ichikawa Y, Kajimoto T (1995) Angew Chem Int Ed Engl 34: 521
43. Fukuda M (1994) Cell surface carbohydrates: cell type specific expression In: Fukuda M, Hindsgaul O (eds) Frontiers in Molecular Biology, Molecular Glycobiology. IRL Press, Oxford, p 1
44. Schachter H (1995) Biosynthesis 2c Glycosyltransferases involved in the synthesis of N-glycan antennae. In: Montreuil J, Vliegenthart JFG, Schachter H (eds) (1995) New Comprehensive Biochemistry Vol 29a, Glycoproteins. Elsevier, Amsterdam, p 153
45. Paulson JC (1989) TIBS 14: 272
46. Brockhausen I (1995) Biosynthesis 3. Biosynthesis of O-glycans of the N-acetylgalactosamine-α-Ser/Thr linkage type. In: Montreuil J, Vliegenthart JFG, Schachter H (eds) (1995) New Comprehensive Biochemistry Vol 29a, Glycoproteins. Elsevier, Amsterdam, p 201
47. Watkins WM (1995) Biosynthesis 5. Molecular basis of t antigenic specificity in the A, B, 0, H and Lewis blood group system. In: Montreuil J, Vliegenthart JFG, Schachter H (eds) (1995) New Comprehensive Biochemistry Vol 29a, Glycoproteins. Elsevier, Amsterdam, p 313
48. Stroud MR, Levery SB, Hakomori S-I (1993) Extended type I glycosphingolipids Lea-Lea (Dimeric Lea) and Lea-Leb as human tumor associated antigens. In: Garegg PJ, Lindberg AA (eds) Carbohydrate antigens. American Chemical Society, New York, p 159
49. Jentoft N (1990) TIBS 15: 291
50. Schachter H, Brockhausen I (1992) The biosynthesis of Serine/Threonine-N-Acetylgalactosamine-Linked carbohydrate moieties. In: Allen HJ, Kisalius EC (eds) Glycoconjugates: composition, structure and function Marcel Dekker, New York, p 263
51. Chai W, Hounsell EF, Cashmore GC, Rosankiewicz JR, Bauer C (1992) Eur J Biochem 203: 257
52. Levery SB, Weiss JB, Salyan ME, Roberts CE, Hakomori S-I (1992) J Biol Chem 267: 5542
53. Makaaru CK, Damian RT, Smith DF, Cummings RD (1992) J Biol Chem 267: 2251
54. Neeleman AP (1996) PhD-thesis, Vrije Universiteit Amsterdam
55. Muramatsu T (1988) Biochmie 70: 1587
56. Drickamer K (1994) Molecular structure of animal lectins. In: Fukuda M, Hindsgaul O (eds) Frontiers in Molecular Biology, Molecular Glycobiology. IRL Press, Oxford, p 53
57. Lasky LA (1992) Science 258: 964
58. Lowe JB (1994) Carbohydrate recognition in cell-cell interaction. In: Fukuda M, Hindsgaul O (eds) Frontiers in Molecular Biology, Molecular Glycobiology. IRL Press, Oxford, p 163
59. Ogawa J-I, Inoue H, Koide S (1996) Cancer Res 56: 325
60. Mulligan MS, Paulson JC, De Frees S, Zheng Z-L, Ward P (1993) Nature 364: 149
61. Messner P, Sleytr UB (1991) Glycobiology 1: 545
62. Sumper M, Wiel FT (1995) Bacterial glycoproteins. In: Montreuil J, Vliegenthart JFG, Schachter H (eds) (1995) New Comprehensive Biochemistry, Vol 29a, Glycoproteins. Elsevier, Amsterdam, p 455
63. Hartmann E, Messner P, Almeier G, König H (1993) J Bacteriol 175: 4515

64. Hartmann E, König H (1989) Arch Microbiol 151: 274
65. König H, Kandler O, Hammes W (1989) Can J Microbiol 35: 176
66. Hartmann E, König H (1990) Naturwissenschaften 77: 472
67. Rietschel ET, Brade L, Schade FU, Seydel U, Zähringer U, Mamat U, Schmidt G, Ulmer A-J, Loppnow H, Flad H-D, di Padova F, Schreier MH, Brade H (1993) Imun Infekt 21: 26
68. Zähringer U, Lindner B, Rietschel ET (1994) Adv Carbohydr Chem Biochem 50: 211
69. Jiang X-M, Neal B, Lee SJ, Romana LK, Reeves PR (1991) Mol Microbiol 5: 695
70. Aspinall GO, Monteiro MA, Pang H, Walsh EJ, Moran AP (1994) Carbohydr Lett 1: 151
71. Aspinall GO, Monteiro MA (1996) Biochemistry 35: 2498
72. Aspinall GO, Monteiro MA, Pang H, Walsh EJ, Moran AP (1996) Biochemistry 35: 2489
73. Chan NWC, Stangier K, Sherburne R, Taylor DE, Zhang Y, Dovichi NJ, Palcic M (1995) Glycobiology 5: 683
74. Seno ET, Hutchinson CR (1986) The biosynthesis of tylosin and erythromycin: Model systems for the studies of the genetics and biochemistry of antibiotic formation. In: Queener SW, Day LE (eds): The Bacteria Volume IX Antibiotic-Producing Streptomyces. Academic Press, Orlando, p 231
75. Okuda T, Ito Y (1982) Biosynthesis and mutasynthesis of aminoglycoside antibiotics. In: Umezawa H, Hooper IR (eds) Aminoglycoside Antibiotics. Springer, Berlin Heidelberg New York, p 111
76. Omura S (1992) The search for bioactive compounds from microorganisms. Springer Berlin Heidelberg New York
77. Peschke U, Schmidt H, Zhang H-Z, Piepersberg W (1995) Mol Microbiol 16: 1137
78. Sinnott ML (1990) Chem Rev 90: 1171
79. Withers SG, Street IP (1988) J Am Chem Soc 110: 8551
80. Withers SG, Warren AJ, Street IP, Rupitz K, Kempton JB, Aebersold R (1990) J Am Chem Soc 112: 5887
81. Withers SG, Aebersold R (1995) Protein Science 4: 361
82. Nilsson KGI (1988) TIBTECH 6: 256
83. Larsson P-O, Hedbys L, Svensson S, Mosbach K (1987) Meth. Enzymol 136: 230
84. Nilsson KGI (1987) Carbohydr Res 167: 95
85. Johansson E, Hedbys L, Larsson P-O, Mosbach K, Gunarsson A (1986) Biotechnol Lett 8: 421
86. Ajisaka K, Nishida H, Fujimoto H (1987) Biotechnol Lett 9: 243
87. Hedbys L, Larsson P-O, Mosbach K (1984) Biochem Biophys Res Commun 123: 8
88. Herrmann GF, Kragl U, Wandrey C (1993) Angew Chem Int Ed Engl 32: 1342
89. Sakai K, Katsumi R, Ohi H, Usui T, Ishido Y (1992) J Carbohydr Chem 11: 553
90. Usui T, Kubota S, Ohi H (1993) Carbohydr Res 244: 315
91. Ajisaka K, Nishida H, Fujimoto H (1987) Biotechnol Lett 9: 387
92. Nilsson KGI (1988) Carbohydr Res 180: 53
93. Crout DH, Howarth OW, Singh S, Swoboda BEP, Critchley P, Gibson WT (1991) J Chem Soc Chem Commun: 1550
94. Crout DHG, MacManus DA, Ricca J-M, Critchley P, Gibson WT (1992) Pure Appl Chem 64: 1079
95. Crout DHG, Singh S, Swoboda BEP, Critchley P, Gibson WT (1992) J Chem Soc Chem Commun: 704
96. Wong C-H, Whitesides GM (1994) Enzymes in Organic Synthesis. Elsevier Science, Oxford
97. Kren V, Thiem J (1995) Angew Chem Int Ed Engl 34: 893
98. Leloir L (1971) Science 172: 1299
99. Beyer TA, Sadler JE, Rearick JI, Paulson JC, Hill RL (1981) Glycosyltransferases and their use in assessing oligosaccharide structure and structure-function relationships. In: Meister A (ed) Advances in Enzymology, Related Molecular Biology Vol 52. Wiley, New York, p 24
100. Kleene R, Berger E (1993) Biochim Biophys Acta 1154: 283
101. Palcic MM (1994) Meth Enzymol 230: 300
102. Field MC, Wainwright LJ (1995) Glycobiology 5: 463
103. Wainwright LJ, Field MC (1996) Glycobiology 6: 5
104. Dinter A, Berger EG (1995) The regulation of cell- and tissue-specific expression of glycans by glycosyltransferases. In: Alavi A, Axford JS (eds) Glycoimmunology. Plenum Press, New York, p 53
105. Paulson JC, Colley KJ (1989) J Biol Chem 264: 17615
106. Nilsson T, Lucocq JM, Mackay D, Warren G (1991) EMBO J 10: 3567

107. Teasdale RD, D'Agostaro G, Gleeson PA (1992) J Biol Chem 267: 4084
108. Aoki D, Lee N, Yamaguchi N, Dubois C, Fukuda MN (1992) Proc Natl Acad Sci USA 89: 4319
109. Masibay AS, Balaji PV, Boeggeman EE, Qasba PK (1993) J Biol Chem 268: 9908
110. Russo RN, Shaper NL, Taatjes DJ, Shaper JH (1992) J Biol Chem 267: 9241
111. Munro S (1991) EMBO J 10: 3577
112. Burke J, Pettitt JM, Schachter H, Sarkar M, Gleeson PA (1992) J Biol Chem 267: 24433
113. Tang BL, Wong SH, Low SH, Hong W (1992) J Biol Chem 267: 10122
114. Elhammer A, Kornfeld S (1986) J Biol Chem 261: 5249
115. Nishikawa Y, Pegg W, Paulsen H, Schachter H (1988) J Biol Chem 263: 8270
116. Sarnesto A, Köhlin T, Hindsgaul O, Thurin J, Blaszczyk-Thurin M (1992) J Biol Chem 267: 2737
117. Sarnesto A, Köhlin T, Hindsgaul O, Vogele K, Blaszczyk-Thurin M, Thurin J (1992) J Biol Chem 267: 2745
118. Sugiura M, Kawasaki T, Yamashina I (1982) J Biol Chem 257: 9501
119. Weinstein J, de Souza-e-Silva U, Paulson JC (1982) J Biol Chem 257: 13845
120. Aoki D, Appert HE, Johnson D, Wong SS, Fukuda MN (1990) EMBO J 9: 3171
121. Wang Y, Wong SS, Fukuda MN, Zu H, Liu Z, Tang Q, ApperT HE (1994) Biochem Biophys Res Commun 204: 701
122. Zu H, Fukuda MN, Wong SS, Wang Y, Liu Z, Tang Q, Appert HE (1995) Biochem Biophys Res Commun 206: 362
123. Wen DX, Livingston BD, Medzihradszky KF, Kelm S, Burlingame AL, Paulson JC (1992) J Biol Chem 267: 21011
124. Drickamer K (1993) Glycobiology 3: 2
125. Datta AK, Paulson JC (1995) J Biol Chem 270: 1497
126. Tsuji S (1995) RIKEN Rev 8: 5
127. Holmes EH, Xu Z, Sherwood L, Macher BA (1995) J Biol Chem 270: 8145
128. Holmes EH (1992) Arch Biochem Biophys 296: 562
129. Liu D, Haase AM, Lindqvist L, Lindberg AA, Reeves PR (1993) J Bacteriol 175: 3408
130. David S, Augé C, Gautheron C (1991) Adv Carbohydr Chem Biochem 49: 175
131. Ichikawa Y, Look GC, Wong C-H (1992) Anal Biochem 202: 215
132. Palcic MM, Hindsgaul O (1996) Trends in Glycosci Glycotechnol, 8: 37
133. Thiem J (1995) FEMS Microbiol Rev 16: 193
134. Waldmann H (1995) Enzymatic synthesis of O-glycosides. VCH, Weinheim
135. Baisch G, Öhrlein R, Ernst B (1996) Bioorg Med Chem Letters 6: 749
136. Baisch G, Öhrlein R, Streiff M, Ernst B (1996) Bioorg Med Chem Letters 6: 755
137. Baisch G, Öhrlein R, Katapodis A, Ernst B (1996) Bioorg Med Chem Letters 6: 759
138. Borsig L, Ivanov SI, Herrmann GF, Kragl U, Wandrey C, Berger EG (1995) Biochem Biophys Res Commun 210: 14
139. Herrmann GF, Wang P, Shen G-J, Garcia-Junceda E, Khan S, Matta KL, Wong C-H (1994) J Org Chem 59: 6356
140. Herrmann GF, Krezdorn C, Malissard M, Kleene R, Paschold H, Weuster-Botz D, Kragl U, Berger EG, Wandrey C (1995) Prot Expr Purif 6: 72
141. Nakazawa K, Furukawa K, Narimatsu H, Kobata A (1993) J Biochem 113: 747
142. Krezdorn CH, Watzele G, Kleene RB, Ivanov SX, Berger EG (1993) Eur J Biochem 212: 113
143. Larsen RD, Rajan VP, Ruff MM, Kukowska-Latallo J, Cummings R, Lowe JB (1989) Proc Natl Acad Sci USA 86: 8227
144. Joziasse DH, Shaper NL, Van den Eijnden DH, Van der Spoel A, Shaper JH (1990) Eur J Biochem 191: 75
145. Seto NOL, Palcic MM, Hindsgaul O, Bundle DR, Narang S (1995) Eur J Biochem 234: 323
146. Herrmann GF, Wang P, Shen G-J, Wong C-H (1994) Angew Chem Int Ed Engl 33: 1241
147. Williams MA, Kitagawa H, Datta AK, Paulson JC, Jamieson JC (1995) Glycoconjugate J 12: 755
148. Sarkar M, Schachter H (1992) Glycobiology 2: 483
149. Sarkar M (1994) Glycoconjugate J 11: 204
150. White T, Bennett EP, Takio K, Sorensen T, Bonding N, Clausen H (1995) J Biol Chem 270: 24156
151. Homa FL, Baker CA, Thomsen DR, Elhammer AP (1995) Prot Expr Purif 6: 141
152. Kukowska-Latallo JF, Larsen RD, Nair RP, Lowe JB (1990) Genes Dev 4: 1288

153. Gersten KM, Natsuka S, Trinchera M, Petryniak B, Kelly RJ, Hiraiwa N, Jenkins NA; Gilbert DJ, Copeland NG, Lowe JB (1995) J Biol Chem 270: 25047
154. De Vries T, Srnka CA, Palcic MM, Swiedler SJ, Van den Eijnden DH (1995) J Biol Chem 270: 8712
155. Feingold DS, Barber GA (1990) Nucleotide sugars. In: Dey PM, Harborne JB (eds) Methods in Plant Biochemistry Vol 2. Academic Press, New York, p 39
156. Gabriel O, van Lenten L (1978) The interconversion of monosaccharides. In: Manners DJ (ed) International Review of Biochemistry Biochemistry of Carbohydrates II Vol 16. University Park Press, Baltimore, p 1
157. Stoddart RW (1984) Sugar nucleotides and cyclitols. In: Stoddart RW The biosynthesis of polysaccharides. Croom Helm, London, p 27
158. Adelhorst K, Whitesides GM (1993) Carbohydr Res 242: 69
159. Arlt M, Hindsgaul O (1995) J Org Chem 60: 14
160. Ichikawa Y, Sim MM, Wong C-H (1992) J Org Chem 57: 2943
161. Schmidt RR, Wegmann B, Jung K-H (1991) Liebigs Ann Chem: 121
162. Müller T, Schmidt RR (1995) Angew Chem Int Ed Engl 34: 1328
163. Srivastava G, Hindsgaul O, Palcic MM (1993) Carbohydr Res 245: 137
164. Gokhale UB, Hindsgaul O, Palcic MM (1990) Can J Chem 68: 1063
165. Lindhorst TK, Thiem J (1990) Liebigs Ann Chem: 1237
166. Lindhorst TK, Thiem J (1991) Carbohydr Res 209: 119
167. Leon B, Lindhorst TK, Rieks-Everdiking A, Klaffke W (1994) Synthesis July: 689
168. Hällgren C, Hindsgaul O (1995) J Carbohydr Chem 14: 453
169. Kochetkov K, Shibaev VN (1973) Adv Carbohydr Chem Biochem 28: 304
170. Klaffke W (1994) Carbohydrates in Europe 10: 9
171. Williams N, Wander J (1980) Deoxy and branched-chain sugars. In: Pigman W, Horton D (eds) The Carbohydrates Chemistry, Biochemistry Vol 1B. Academic Press, New York, p 761
172. Beck E (1982) Branched-chain sugars. In: Loewus FA, Tanner W (eds) Encyclopedia of plant physiology New Series Volume 13A Carbohydrates I Intracellular Carbohydrates. Springer, Berlin Heidelberg New York, p 124
173. Flowers HM (1981) Adv Carbohydr Chem Biochem 39: 279
174. Shibaev VN (1978) Pure Appl Chem 50: 1421
175. Shibaev VN, Danilov LL, Druzhinina TN, Gogolashwili LM, Maltsev SD (1982) FEBS Lett 139: 177
176. Ashwell G, Volk WA (1965) J Biol Chem 240: 4549
177. Aspinall GO (1981) Constitution of cell wall polysaccharides. In: Tanner W, Loewus FA (eds) Plant carbohydrates I Intracellular carbohydrates Vol 13 B. Springer, Berlin Heidelberg New York, p 3
178. Barber GA (1968) Biochim Biophys Acta: 68
179. Barber GA (1962) Biochem Biophys Res Commun 8: 204
180. Barber GA (1962) Biochemistry 1: 463
181. Beevers L (1982) Amino sugars – Plants and fungi. In: Loewus FA, Tanner W (eds) Encyclopedia of plant physiology New Series Volume 13A Carbohydrates I Intracellular Carbohydrates. Springer, Berlin Heidelberg New York, p 103
182. Bevill RD (1968) Biochem Biophys Res Commun 30: 595
183. Broschat KO, Chang S, Serif GS (1985) Eur J Biochem 153: 397
184. Chang S, Duerr B, Serif GS (1988) J Biol Chem 263: 1693
185. Crawley SC, Hindsgaul O, Ratcliffe RM, Lamontagne LR (1989) Carbohydr Res 193: 249
186. De Vries T, Van den Eijnden DH (1992) Histochem J 24: 761
187. Dumas DP, Ichikawa Y, Wong C-H, Lowe JB, Nair RP (1991) Bioorg Med Chem Letters 1: 425
188. Elbein AD, Heath EC (1965) J Biol Chem 240: 1926
189. Feingold DS (1982) Aldo (and Keto) hexoses and uronic acids. In: Loewus FA, Tanner W (eds) Encyclopedia of plant physiology New Series Volume 13A Carbohydrates I Intracellular Carbohydrates. Springer, Berlin Heidelberg New York, p 3
190. Floss HG, Beale JM (1989) Angew Chem Int Ed Engl 28: 146
191. Franz G (1982) Glycosylation of heterosides (glycosides). In: Loewus FA, Tanner W (eds) Encyclopedia of plant physiology New Series Volume 13A Carbohydrates I Intracellular Carbohydrates. Springer, Berlin Heidelberg New York, p 384
192. Gabriel O (1973) Adv Chem Ser 17: 387

193. Gaugler RW, Gabriel O (1973) J Biol Chem 248: 6041
194. Gilbert JM, Matsuhashi M, Strominger JL (1965) J Biol Chem 240: 1305
195. Ginsburg V (1964) Adv Enzymol 26: 35
196. Ginsburg V (1961) J Biol Chem 236: 2389
197. Ginsburg V (1960) J Biol Chem 235: 2196
198. Glaser L, Zarkowsky H, Ward L (1972) Meth Enzymol 28: 446
199. Glaser L, Zarkowsky H (1971) Enzymes 5: 465
200. Glaser L, Kornfeld S (1961) J Biol Chem 236: 1795
201. Gonzalez-Porque P, Strominger JL (1972) J Biol Chem 247: 6748
202. Gräfe U (1992) Biochemie der Antibiotika Struktur-Biosynthese-Wirkmechanismus. Spektrum, Heidelberg
203. Grisebach H (1978) Adv Carbohydr Chem Biochem 35: 81
204. Hasegawa T, Kakushima M, Hatori M, Aburaki S, Kakinuma S, Furumai T, Oki T (1993) J Antibiotics 46: 598
205. Holmes EH (1993) Glycobiology 3: 77
206. Jarvis BW, Hutchinson CR (1994) Arch Biochem Biophys 308: 175
207. Kennedy JF, White CA (1983) Bioactive carbohydrates in chemistry biochemistry, biology. Ellis Horwood, Chichester
208. Liao T-H, Barer GA (1972) Biochim Biophys Acta 276: 85
209. Markovitz A (1964) J Biol Chem 239: 2091
210. Markovitz A (1963) Federation Proc 22: 464
211. Martin JF (1984) Biosynthesis, regulation, and genetics of polyene macrolide antibiotics. In: Omura S (ed) Macrolide antibiotics Chemistry Biology, and Practice Academic Press, Orlando, 405
212. Marumo K, Lindqvist L, Verma A, Weintraub A, Reeves PR, Lindberg AA (1992) Eur J Biochem 204: 539
213. Matsuhashi S, Strominger JL (1967) J Biol Chem 242: 3494
214. Matsuhashi M, Strominger JL (1966) J Biol Chem 241: 4738
215. Matsuhashi S, Strominger JL (1966) Meth Enzymol 8: 310
216. Matsuhashi M, Strominger JL (1964) J Biol Chem 239: 2454
217. Matsuhashi M (1963) Federation Proc 22: 465
218. Melo A, Elliott WH, Glaser L (1968) J Biol Chem 243: 1467
219. Melo A, Glaser L (1968) J Biol Chem 243: 1475
220. Oerskov F, Oerskov I, Jann B, Jann K, Müller-Seitz E, Westphal O (1967) Acta path et microbiol scandinav 71: 339
221. Ohashi H, Matsuhashi M, Matsuhashi S (1971) J Biol Chem 246: 2325
222. Okazaki R, Okazaki T, Strominger JL, Michelson AM (1962) J Biol Chem 237: 3014
223. Pape H, Brillinger GU (1973) Arch Mikrobiol 88: 25
224. Pazur JH, Shuey EW (1961) J Biol Chem 236: 1780
225. Percival E (1963) The monodsaccharides. In: Florkin M, Stotz EH (eds) Comprehensive biochemistry Carbohydrates Vol 5. Elsevier Amsterdam, 1
226. Reeves PR (1993) TIG 9: 17
227. Rubenstein PA, Strominger JL (1974) J Biol Chem 249: 3782
228. Selvendran RR, O'Neill MA (1982) Plant glycoproteins. In: Loewus FA, Tanner W (eds) Encyclopedia of plant physiology New Series Volume 13A Carbohydrates I Intracellular Carbohydrates. Springer, Berlin Heidelberg New York, p 515
229. Sutherland IW (1990) Biotechnology of microbial exopolysaccharides. Cambridge University Press, Cambridge New York
230. Thompson MW, Strohl WR, Floss HG (1992) J Gen Microbiol 138: 779
231. Thorson JS, Liu H-W (1993) J Am Chem Soc 115: 7539
232. Thorson JS, Lo SF, Liu H-W (1993) J Am Chem Soc 115: 6993
233. Thorson JS, Oh E, Liu H-W (1992) J Amer Chem Soc 114: 6941
234. Tonn SJ, Gander JE (1979) Ann Rev Microbiol 33: 169
235. Vanek Z, Majer J (1967) Macrolide antibiotics. In: Gottlieb D, Shaw PD (eds) Antibiotics Vol 2. Springer, New York Heidelberg 154
236. Vara JA, Hutchinson CR (1988) J Biol Chem 263: 14992
237. Volk WA, Ashwell G (1963) Biochem Biophys Res Commun 12: 116
238. Wahl HP, Grisebach H (1979) Biochim Biophys Acta 568: 243
239. Weigel TM, Liu L-D, Liu H-W (1992) Biochemistry 31: 2129

240. Weigel TM, Miller VP, Liu H-W (1992) Biochemistry 31: 2140
241. Winkler NW, Markovitz A (1971) J Biol Chem 246: 5868
242. Wong C-H, Dumas DP, Ichikawa Y, Koseki K, Danishefsky SJ, Weston BW, Lowe JB (1992)
     J Amer Chem Soc 114: 7321
243. Yamamoto K, Katayama I, Onoda Y, Inami M, Kumagai H, Tochikura T (1993) Arch
     Biochem Biophys 300: 694
244. Yamamoto K, Maruyama T, Kumagai H, Tochikura T, Seno T, Yamaguchi H (1984) Agric
     Biol Chem 48: 823
245. Ishihara H, Massaro DJ, Heath EC (1968) J Biol Chem 243: 1103
246. Ishihara H, Heath EC (1968) J Biol Chem 243: 1110
247. Butler W, Serif GS (1985) Biochim Biophys Acta 829: 238
248. Isselbacher K (1958) J Biol Chem 232: 429
249. Abraham HD, Howell RR (1969) J Biol Chem 244: 545
250. Kingsley DM, Kozarsky KF, Hobbie L, Krieger M (1986) Cell 44: 749
251. Szumilo T, Zeng Y, Pastuszak I, Drake R, Szumilo H, Elbein AD (1996) J Biol Chem 271:
     13147
252. Gross KC, Pharr DM (1982) Plant Physiol 69: 117
253. Dressler K, Bliedlingmaier S, Grossberger H, Kemmer J, Nölle U, Rodmanis-Blumer A, Hess
     D (1982) Z Pflanzenphysiologie 107: 409
254. Maretzki A, Thom M (1978) Plant Physiol 61: 544
255. Schachter H, Ishihara H, Heath EC (1972) Meth Enzymol 28: 285
256. Ichikawa Y, Wang R, Wong C-H (1994) Meth Enzymol 247: 107
257. Toone EJ, Whitesides GM (1991) Am Chem Soc Sympos Ser 466: 1
258. Unverzagt C, Paulson JC (1990) J Amer Chem Soc 112: 9308
259. Wong C-H, Haynie SL, Whitesides GM (1982) J Org Chem 47: 5416
260. Kleczkowski LA (1994) Phytochemistry 37: 1507
261. Turnquist RL, Hansen RG (1973) Uridine diphosphoryl glucose pyrophosphorylase. In: Boyer
     PD (ed) The Enzymes Vol VIII. Academic Press, New York, 51
262. Wong C-H, Dreckhammer DG, Sweers HM (1988) Am Chem Soc Sympos Ser 374: 30
263. Heidlas JE, Lees WJ, Pale P, Whitesides GM (1992) J Org Chem 57: 146
264. Ropp PA, Cheng P-W (1990) Anal Biochem 187: 104
265. Drake RR, Evans RK, Wolf MJ, Haley BE (1989) J Biol Chem 264: 11928
266. Konishi Y, Tanizawa K, Muroya S, Fukui T (1993) J Biochem 114: 61
267. Katsube T, Kazuta Y, Tanizawa K, Fukui T (1991) Biochemistry 30: 8546
268. Hossain SA, Tanizawa K, Kazuta Y, Fukui T (1994) J Biochem 115: 965
269. Elling L, Kula M-R (1994) J Biotechnol 34: 157
270. Elling L (1996) Phytochemistry 42: 955
271. Herrmann GF, Elling L, Berger EG, Wandrey C (1995) Bioorg Med Chem 5: 673
272. Zervosen A, Stein A, Adrian H, Elling L (1996) Tetrahedron 52: 2395
273. Elling L, Grothus M, Kula M-R (1993) Glycobiology 3: 349
274. Elling L, Kula M-R (1995) Enzyme Microb Technol 17: 929
275. Zervosen A, Elling L (1996) J Am Chem Soc 118: 1836
276. Hokke CH, Zervosen A, Elling L, Joziasse DH, Van den Eijnden (1996) Glycoconjugate J 13:
     687
277. Korf U, Thimm J, Thiem J (1991) Synlett April: 313
278. Mengin-Lecreulx D, Van Heijenoort J (1993) J Bacteriol 175: 6150
279. Lang L, Kornfeld S (1984) Anal Biochem 140: 264
280. Look GC, Ichikawa Y, Shen G-S, Cheng P-W, Wong C-H (1993) J Org Chem 58: 4326
281. De Luca C, Lansing M, Martini I, Crescenzi F, Shen G-S, O'Regan M, Wong C-H (1995) J Am
     Chem Soc 117: 5869
282. De Luca C, Lansing M, Crescenzi F, Martini I, Shen G-J, O'Regan M, Wong C-H (1996)
     Bioorg Med Chem 4: 131
283. Lindqvist L, Kaiser R, Reeves PR, Lindberg AA (1993) Eur J Biochem 211: 763
284. Zervosen A, Elling L, Kula M-R (1994) Angew Chem Int Ed Engl 33: 571
285. Lindqvist L, Kaiser R, Reeves PR, Lindberg AA (1994) J Biol Chem 269: 122
286. Ballicora MA, Laughlin MJ, Fu Y, Okita TW, Barry GF, Preiss J (1995) Plant Physiol 109: 245
287. Okita TW, Rodriguez RL, Preiss J (1981) J Biol Chem 256: 6944
288. Ichikawa Y, Liu JL-C, Shen G-J, Wong C-H (1991) J Amer Chem Soc 113: 6300
289. Ichikawa Y, Shen G-J, Wong C-H (1991) J Amer Chem Soc 113: 4698

290. Ichikawa Y, Lin Y-C, Dumas DP, Shen G-J, Garcia-Junceda E, Williams MA, Bayer R, Ketcham C, Walker LE, Paulson JC, Wong C-H (1992) J Amer Chem Soc 114: 9283
291. David S, Augé C (1987) Pure Appl Chem 59: 1501
292. Simon ES, Toone EJ, Ostroff G, Bednarski M, Whitesides GM (1989) Meth Enzymol 179: 275
293. Augé C, Fernandez R, Gautheron C (1990) Carbohydr Res 200: 257
294. Thiem J, Treder W (1986) Angew Chem Int Ed Engl 25: 1100
295. Thiem J, Stangier P (1990) Liebigs Ann Chem: 1101
296. Kittelmann M, Klein T, Kragl U, Wandrey C, Ghisalba O (1995) Appl Microbiol Biotechnol 44: 59
297. Shames SL, Simon ES, Christopher CW, Schmid W, Whitesides GM, Yang L-L (1991) Glycobiology 1: 187
298. Shen G-J, Liu JL-C, Wong C-H (1992) Biocatalysis 6: 31
299. Augé C, Gautheron C (1988) Tetrahedron Lett 29: 789
300. Brossmer R, Gross HJ (1994) Meth Enzymol 247: 153
301. Brossmer R, Gross HJ (1994) Meth Enzymol 247: 177
302. Sugai T, Lin C-H, Shen G-J, Wong C-H (1995) Bioorg Med Chem 3: 313
303. Grier TJ, Rasmussen JR (1982) Anal Biochem 127: 100
304. Klaffke W (1995) Carbohydr Res 266: 285
305. Leon B, Lindhorst TK, Rieks-Everdiking A, Klaffke W (1994) Synthesis July: 689
306. McDowell W, Grier TJ, Rasmussen JR, Schwarz RT (1987) Biochem J 248: 523
307. Pallanca JE, Turner NJ (1993) J Chem Soc Perkin Trans 1: 3017
308. Simon ES, Grabowski S, Whitesides GM (1990) J Org Chem 55: 1834
309. Szumilo T, Drake RR, York JL, Elbein AD (1993) J Biol Chem 268: 17943
310. Sa-Correira I, Darzins A, Wang S-K, Berry A, Chakrabarty AM (1987) J Bacteriol 169: 3224
311. Elling L, Ritter JE, Verseck S (1996) Glycobiology 6: 591
312. Stiller R, Thiem J (1992) Liebigs Ann Chem 5: 467
313. Thiem J, Wiemann T (1991) Angew Chem Int Ed Engl 30: 1163
314. Augé C, David S, Mathieu C, Gautheron C (1984) Tetrahedron Lett 25: 1467
315. Bauer AJ, Rayment I, Frey PA, Holden HM (1991) Proteins: Structure Function Genetics 9: 135
316. Dörmann P, Benning C (1996) Arch Biochem Biophys 327: 27
317. Wong C-H, Wang R, Ichikawa Y (1992) J Org Chem 57: 4343
318. Heidlas JE, Lees WJ, Whitesides GM (1992) J Org Chem 57: 152
319. Bülter T, Wandrey C, Elling L (1996) unpublished results
320. Nassau PM, Martin SL, Brown RE, Weston A, Monsey D, McNeil MR, Duncan K (1996) J Bacteriol 178: 1047
321. Gygax D, Spies P, Winkler T, Pfaar U (1991) Tetrahedron 47: 5119
322. Drake RR, Zimniak P, Haley BE, Lester R, Elbein AD, Radominska A (1991) J Biol Chem 266: 23257
323. Toone EJ, Simon ES, Whitesides GM (1991) J Org Chem 56: 5603
324. Liljebjelke K, Adolphson R, Baker K, Doon RL, Mohnen D (1995) Anal Biochem 225: 296
325. Kyossev ZN, Drake RR, Kyosseva S, Elbein AD (1995) Eur J Biochem 228: 109
326. Gubler M, Appoldt Y, Keck W (1996) J Bacteriol 178: 906
327. Marquardt JL, Siegele DA, Kolter R, Walsh CT (1992) J Bacteriol 174: 5748
328. Brown ED, Marquardt JL, Lee JP, Walsh CT, Anderson KS (1994) Biochemistry 33: 10638
329. Benson TE, Marquardt JL, Marquardt AC, Etzkorn FA, Walsh CT (1993) Biochemistry 32: 2024
330. Tayeh MA, Dotson GD, Clemens JC, Woodard RW (1995) Prot Expr Purif 6: 757
331. Stein A, Kula M-R, Elling L, Verseck S, Klaffke W (1995) Angew Chemie Int Ed Engl 34: 1748
332. Stein A (1995) PhD thesis, Heinrich-Heine-University Düsseldorf, Düsseldorf, Germany
333. Lindqvist L, Schweda KH, Reeves PR, Lindberg AA (1994) Eur J Biochem 225: 863
334. Tsuboi KK, Fukunga K, Petricciani JC (1969) J Biol Chem 244: 1008
335. Hsiao P-H, Su J-C, Sung H-Y (1981) Proc Natl Sci Counc B ROC 5: 31
336. Ritter JE, Berlin C, Elling L (1996) Anal Biochem 234: 74
337. Kleczkowski LA, Viland P, Olsen O-A (1993) Z Naturforsch 48c: 457
338. Lehle L, Tanner W (1995) Protein glycosylation in yeast. In: Montreuil J, Vliegenthart JFG, Schachter H (eds) (1995) New Comprehensive Biochemistry Vol 29a, Glycoproteins. Elsevier, Amsterdam, 475
339. Flitsch SL, Taylor JP, Turner NJ (1991) J Chem Soc Chem Commun: 382

340. Wang P, Shen G-S, Wang Y-F, Ichikawa Y, Wong C-H (1993) J Org Chem 58: 3985
341. Kornfeld RH, Ginsburg V (1966) Biochim Biophys Acta 117: 79
342. Munch-Petersen A (1956) Acta Chem Scand 10: 928
343. Munch-Petersen A (1962) Meth Enzymol 5: 171
344. Braell WA, Tyo MA, Krag SS, Robins PW (1976) Anal Biochem 74: 484
345. Niggemann J, Thiem J (1992) Liebigs Ann Chem 5: 535
346. Preiss J, Greenberg E (1972) Meth Enzymol 28: 281
347. Frey PA (1987) Complex pyridine-dependent transformations. In: Dolphin D, Poulson R, Avamovic O (eds): Pyridine nucleotide coenzymes Chemical biochemical, medical aspects Vol 2B. Wiley, New York, 462
348. Glaser L (1959) J Biol Chem 234: 2801
349. Echt CS, Chourey PS (1985) Plant Physiol 79: 530
350. Copeland L (1990) Enzymes of surcose metabolism In: Lea PJ (ed) Methods in Plant Biochemistry Vol 3. Academic Press, New York, 73
351. Neufeld EF, Hassid WZ (1963) Adv Carbohydr Chem 18: 309
352. Avigad G (1982) Surcose and other disaccharides. In: Loewus FA, Tanner W (eds) Encyclopedia of plant physiology New Series Volume 13A Carbohydrates I Intracellular Carbohydrates. Springer-Verlag, Berlin, 217
353. Elling L, Kula M-R (1993) J Biotechnol 29: 277
354. Elling L, Grothus M, Zervosen A, Péus M, Helfer A, Stein A, Adrian H, Kula M-R (1995) Biotechnol Appl Biochem 21: 29
355. Elling L (1995) Glycobiology 5: 201
356. Peters J, Brockamp H-P, Minuth T, Grothus M, Steigel A, Kula M-R, Elling L (1993) Tetrahedron: Asymmetry 4: 1173
357. Grothus M, Steigel A, Kula M-R, Elling L (1994) Carbohydrate Letters 1: 83
358. Mengin-Lecreulx D, van Heijenoort J (1994) J Bacteriol 176: 5788
359. Drake RR, Elbein AD (1992) Glycobiology 2: 279
360. Radominska A, Drake RR (1994) Meth Enzymol 230: 330
361. Fukusawa T, Obonai K, Segawa T, Nogi Y (1980) J Biol Chem 255: 2705
362. Nelsestuen GL, Kirkwood S (1971) J Biol Chem 246: 7533
363. Nunez HA, Barker R (1976) Biochemistry 15: 3843

New Alcohol Dehydrogenases for the Synthesis of Chiral Compounds

Werner Hummel
Institut für Enzymtechnologie, der Heinrich-Heine-Universität,
Forschungszentrum Jülich, Postfach 2050, D-52404 Jülich, Germany

Dedicated to Professor Dr. Maria-Regina Kula on the occasion of her 60th birthday

The enantioselective reduction of carbonyl groups is of interest for the production of various chiral compounds such as hydroxy acids, amino acids, hydroxy esters, or alcohols. Such products have high economic value and are most interesting as additives for food and feed or as building blocks for organic synthesis. Enzymatic reactions or biotransformations with whole cells (growing or resting) for this purpose are described. Although conversions with whole cells are advantageous with respect to saving expensive isolation of the desired enzymes, the products often lack high enantiomeric excess and the process results in low time-space-yield. For the synthesis of chiral alcohols, only lab-scale syntheses with commercially available alcohol dehydrogenases have been described yet. However, most of these enzymes are of limited use for technical applications because they lack substrate specificity, stability (yeast ADH) or enantioselectivity (*Thermoanaerobium brockii* ADH). Furthermore, all enzymes so far described are forming (*S*)-alcohols. Quite recently, we found and characterized several new bacterial alcohol dehydrogenases, which are suited for the preparation of chiral alcohols as well as for hydroxy esters in technical scale. Remarkably, of all these novel ADHs the (*R*)-specific enzymes were found in strains of the genus *Lactobacillus*. Meanwhile, these new enzymes were characterized extensively. Protein data (amino acid sequence, bound cations) confirm

Advances in Biochemical Engineering/
Biotechnology, Vol. 58
Managing Editor: Th. Scheper
© Springer-Verlag Berlin Heidelberg 1997

that these catalysts are novel enzymes. (R)-specific as well as (S)-specific ADHs accept a broad variety of ketones and ketoesters as substrates. The applicability of alcohol dehydrogenases for chiral syntheses as an example for the technical use of coenzyme-dependent enzymes is demonstrated and discussed in this contribution. In particular NAD-dependent enzymes coupled with the coenzyme regeneration by formate dehydrogenase proved to be economically feasible for the production of fine chemicals.

Abbreviations

ADH   = alcohol dehydrogenase
YADH   = yeast alcohol dehydrogenase
HLADH = horse liver alcohol dehydrogenase
TBADH = *Thermoanaerobium brockii* alcohol dehydrogenase
FDH   = formate dehydrogenase

1 Introduction

1.1 Dehydrogenases as Catalysts for the Synthesis of Chiral Compounds

Enzymes are attractive catalysts for organic synthesis. They are highly selective and specific and many problems of chemical synthesis such as isomerization, racemization or rearrangement can be avoided because of the mild enzymatic reaction condition. Particularly for asymmetric syntheses, enzyme catalyzed reactions offer notable advantages for the organic chemist. Considering the six main enzyme groups according to the classification of the International Union of Biochemistry [1] the oxidoreductases, hydrolases and lyases represent the most useful enzymes. Hydrolases and lyases are catalysts which in general do not need additional cofactors or coenzymes which simplify the process engineering and optimization of the reaction. Oxidoreductases require coenzymes which complicates the process engineering and tends to increase the product costs; coenzymes are too expensive to provide them in stoichiometric amounts and efficient steps for their regeneration are required.

During the last years, enzyme-catalyzed syntheses had been developed which demonstrate the applicability of such methods. Lipases and esterases are well established tools to produce chiral materials. Processes using coenzyme-dependent enzymes were developed during the last years for the preparation of optically active hydroxy and amino acids employing commercially available enzymes such as lactate dehydrogenases or new dehydrogenases such as *L*- or *D*-hydroxyisocaproate dehydrogenases or *L*-phenylalanine dehydrogenase (Review [2–4]).

Enzyme-catalyzed syntheses of enantiomeric pure hydroxy acids use lactate dehydrogenases (LDH; E.C. 1.1.1.27) or hydroxyisocaproate dehydrogenases (HicDH). Both kinds of enzymes are available as *D*- or *L*-specific catalysts. *L*-specific [5, 6] LDHs as well as *D*-specific [7–10] LDHs favorably catalyze the reduction of pyruvate, HicDHs (and mandelate dehydrogenase, too) convert keto acids with longer aliphatic or aromatic side chains. These enzymes can be isolated from *Lactobacillus* strains [11–14].

Reductive amination reactions of keto acids are performed with amino acid dehydrogenases. NAD-dependent leucine dehydrogenase from *Bacillus* sp. is of interest for the synthesis of (*S*)-*tert.*-leucine [15–17]]. This chiral compound has found widespread application in asymmetric synthesis and as a building block of biologically active substances. The enzyme can also be used for the chemoenzymatic preparation of (*S*)-hydroxy-valine [18] and unnatural hydrophobic branched-chain (*S*)-amino acids. NAD-dependent *L*-phenylalanine dehydrogenase from *Rhodococcus* sp. [19] has been used for the synthesis of *L*-homophenylalanine ((*S*)-2-Amino-4-phenylbutanoic acid) [9]. These processes with water-soluble substrates and products demonstrate that the use of coenzymes must not be a limiting component, the coenzymes are stable enough for long-term applications and using an efficient coenzyme-regeneration step and an appropriate reaction engineering it is possible to reach sufficient turnover numbers for the coenzyme (significantly more than 100,000 are obtainable) that such reactions become economically feasible.

Chiral alcohols are valuable products mainly as building blocks for pharmaceuticals or agro chemicals or as part of chiral catalysts. Cheap biotransformation methods for the selective reduction of particular ketone compounds are known for many years rather catalyzed by fermentation than with isolated enzymes. Products prepared with whole cells such as baker's yeast often lack high enantioselectivity and there were several attemps to use isolated enzymes. Resolution of racemates with hydrolases are known in some cases but very often the reduction of the prochiral ketone using alcohol dehydrogenases are much more attractive.

1.2 Prerequisites for the Use of Dehydrogenases

Alcohol dehydrogenases are enzymes that are well known from physiological and biochemical studies on the primary metabolism of cells. Several ADHs are commercially available and for some of them such as the ADH from yeast or liver details concerning the structure and reaction mechanism have been elucidated. For preparative applications however they seldom meet the requirements and new enzymes are needed for this field.

Several prerequisites have to be fulfilled to make an efficient application of these enzymes possible:

– Enzymes Should Work with High Selectivity and High Yield

In the use of whole cells severe problems may arise from strain specific activity of intracellular enzymes reacting with the product. For instance, the organism may use the substrate or the product as the carbon source or intracellular esterase activities may influence the yield of hydroxy esters formed by enzymatic reduction of keto esters. These problems may be avoided using isolated enzymes. These potential side reactions also define the purification grade of a technical enzyme sample because the complete separation of the disturbing activities must be ensured.

– Enzymes Should Work with High Enantioselectivity and Predictions About Their Stereospecificity Should be Possible

Although conversions with whole cells save expensive procedures to isolate the desired enzymes, the products however often lack sufficient enantiomeric excess and the predicition of the stereospecificity of the reaction is difficult. In most cases processes with whole cells result only in low space-time-yield.

Because of its availability, baker's yeast is by far the most widely used biological catalyst for enantioselective reduction reactions [20–25]. Actually it was applied to nearly countless syntheses of β-hydroxyesters [26, 27]. The results demonstrate that the enantioselectivity of yeast catalyzed reductions of β-ketoesters depends on the structure of the substrate. Along a homologous series of C_1- to C_{12}-esters of 4-chloroacetoacetic acid, a shift in the stereochemistry of the alcohol produced was observed from the (S)-hydroxyester with an ee = 0.7% using the C_1-(methyl)-group to the (R)-hydroxyester with an ee = 95% containing the octyl-group [25]. Detailed biochemical characterization of the enzymes involved revealed that baker's yeast does possess three enzymes capable to reduce β-ketoesters, none of which however is the commercially available yeast ADH, which is thus obviously not involved in such reactions. Two of the isolated enzymes are (R)- and one is (S)-specific. The substrate specificity of these enzymes shows that in increasing the chain length in the ketone led to contrary reaction rates of the three enzymes, thus producing often a mixture of isomers. Therefore it is not possible to predict the stereospecificity of a reaction using a new substrate or a new yeast strain. Otherwise, the enantioselectivity can be altered either by using mutants [28] or various strains of baker's yeast [29–31] or by inhibiting one or several of these enzymes. Addition of allyl alcohols [32] or methyl vinyl ketone [33] supports the formation of (R)-products, presumably by inhibiting an (S)-specific dehydrogenase, but in some cases, results in low yield [34]. Immobilization of the cells [35, 36] or variation of medium constituents may also influence the ratio of isomers, e.g. addition of ammonium salts (53), thiamine (55) or oxygen supply (51.52).

Using isolated enzymes instead of whole cells, similar problems are to be considered only in a few cases. ADH from *Thermoanaerobium brockii* shows varying enantiomeric excess of the product depending on the structure of the ketone to be reduced. Conversions with this enzyme yield in products with low (20% for the reduction of acetophenone) or high ee value (100% for the reduction of p-Cl-acetophenone). Predictions about the stereospecificity of HLADH catalyzed reductions can be made for simple acyclic substrates applying Prelog's rule [37] and for more complex compounds using the cubic-space model developed by Jones and Jakovac [38].

– Efficient Coenzyme-Regeneration Methods Should be Available

Using whole cells for preparative conversions, the metabolic status of the cells is of importance for the efficiency of the regeneration step. This concerns both the available intracellular concentration of the coenzyme and the kind of regeneration. In particular NADP-dependent enzymes may be limited by the level of this coenzyme taking into consideration the Km-values of the dehydrogenases for the coenzymes and the concentrations which are necessary for maximal activity. From results with yeast strains it is likely that the intracellular concentrations of nicotinamide coenzymes differ in strains of various origins [39].

Isolated oxidoreductases always depend on cofactors for the transfer of electrons. Enzyme groups which are well characterized with respect to their biochemistry are those requiring the nicotinamide coenzymes NAD or NADP, the flavins FAD or FMN and the ortho-quinoids such as pyrroloquinoline quinone (PQQ) or trihydroxy-phenylalanine (TOPA).

Among these cofactor-dependent enzymes, the NAD(P)-dependent are of central interest. Due to their high prices, the required coenzymes cannot be applied in stochiometric amounts but must be efficiently regenerated if such enzymes are to be employed as catalysts for preparative chemistry. Generally there are two approaches possible: the application of a second enzyme, or a second substrate. So far, only the use of formate dehydrogenase from *Candida boidinii* for the regeneration of NADH is demonstrated in technical scale. Alternate concepts are discussed such as the application of metallorganic compounds or a modified formate dehydrogenase obtained by protein engineering.

– Enzymes Should be Available in Technical Scale

Several methods for the production and isolation of extra- and intracellular enzymes from wild-type strains are established during the last years. Highly efficient methods for the scale-up of dehydrogenases have been developed for the enzyme formate dehydrogenase [40, 41]. Alternately, gene cloning can be used successfully to obtain an efficient yield of the desired catalyst.

– Enzymes Should Convert Hydrophobic Substrates at an Efficient Space-Time-Yield

Most of the substrates or products which are interesting for organic chemistry are scarcely soluble in aqueous solutions. Unfortunately, enzymes are often deactivated when used in organic solvents. Successful conversions of the hydrophobic keto compounds thus can demonstrate general reaction principles for enzyme-catalyzed transformations of hydrophobic substrates. Various concepts concerning the application of enzymes for the transformation of hydrophobic compounds have been developed, but none of them have been applied so far in technical scale.

– Enzymes can be Modified by Protein Engineering

In order to overcome limitations derived from the protein structure such as lacking stability, substrate or coenzyme specificity, protein engineering methods can be applied. Some examples are published which describe specific mutations of dehydrogenases.

This article will focus on new developments on the field of NAD(P)-dependent dehydrogenases, in particular on new alcohol dehydrogenases applicable for preparative chemistry. Biochemical properties of dehydrogenases useful for the synthesis of chiral hydroxy esters and alcohols are summarized in this contribution with respect to more recent studies on enantioselective reduction of prochiral ketones and to new enzymes for this purpose. Reviews covering earlier works in this field can be found in [42–44].

2 Alcohol Dehydrogenases

2.1 Screening and Site-Directed Mutagenesis for the Development of New Enzymes

Even though there is a great number of known enzymes depending on nicotinamide coenzymes, it seems sensitive for several reasons to screen for novel catalysts with respect to their technical applications. The growing importance of chiral compounds [45–47] demands new enzymes with new substrate specificities. The commercially available enzymes show limitations such as the insufficient operational long-term stability, lack of activity in organic solvents or limited substrate acceptance. The development of highly efficient methods for the regeneration of coenzymes such NADH or NADPH, however, facilitates syntheses that are inexpensive respecting the cofactor involved.

New or improved catalysts can be obtained by two basically various methods: screening for enzymes among naturally available sources or site-directed mutagenesis of known enzymes. Screening procedures with microorganisms, plants or animal cells represent the traditional way to discover new enzymes. Microorganisms are of particular interest because of their short generation time and the large diversity of metabolic pathways and enzymes involved. Special techniques with enrichment cultures or rapid assay methods supported by automation or miniaturization are useful to detect new enzyme producers. However, there are only a few generally applicable screening methods and only a few review articles available on this field [48–51], because successful screening methods essentially depend in detail on the problem to be solved. One strategy to obtain new or improved enzymes is given in Figure 1.

Literature on microbial physiology and biochemistry may be helpful for the choice of strains to find enzymes for the biotransformation of naturally occurring substances. Screening of enzymes to convert unnatural compounds however usually implies an accidental selection of microorganisms. Actually, even such a preliminary approach gives valuable informations about the occurrence of an enzyme activity.

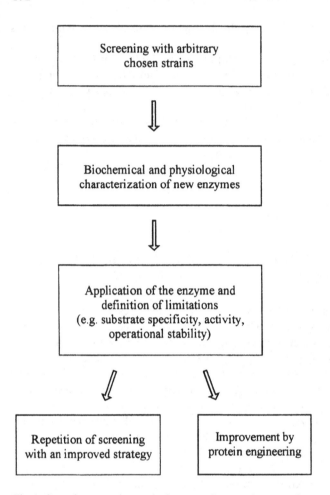

Fig. 1. Screening strategies to obtain new or improved enzymes by screening

Several new enzymes which proved to be useful as technical catalysts were discovered by screening natural habitats. Some examples of NAD(P)-dependent enzymes may demonstrate the various screening procedures: Bacteria with NAD-dependent L-phenylalanine dehydrogenase were isolated from soil samples by providing the culture medium with L-phenylalanine [19, 52–54]. An NAD-independent trimethylamine dehydrogenase could be obtained using enrichment cultivation techniques in a chemostat [55]. A mutant strain of *Rhodobacter sphaeroides* producing a new galactitol dehydrogenase was isolated from a chemostat culture [56]. Strains of *Thermoanaerobium brockii* that produce a thermostable NADP-dependent alcohol dehydrogenase were isolated from a hot spring site in Yellowstone Park [57]. Further thermostable alcohol dehydrogenases could be obtained from a novel strain of *Bacillus stearothermophilus* growing at 70° C [58] and from a novel archaebacterium *Sulfolobus solfataricus* [59, 60]. Enrichment of microorganisms culturing them on selected alcohols as the carbon source resulted in a new NAD-dependent secondary alcohol dehydrogenase with (R)-selectivity for the alcohol [61, 62].

Further improvements to an enzyme can be made by testing related strains in order to find enzyme variants. No general rule exists about the distribution of certain enzymes within microbial groups; some enzymes such as lactate dehydrogenase occur in all organisms of a family (*Lactobacillaceae* in this case), others such as leucine dehydrogenase or alanine dehydrogenase are found in all strains of a genus (*Bacillus*) and some seem to occur only in a single strain of a species. Phenylalanine dehydrogenase for instance occurs in *Rhodococcus erythropolis* and in strains of a new species of *Rhodococcus* [19], but it does not in other *Rhodococcus* species [54]. Obviously, this is not due to the sample preparation or the assay, because the occurrence of phenylalanine dehydrogenase is correlated with the ability to degrade of L-phenylalanine and the inactive species of *Rhodococcus* were found to be unable to catabolize phenylalanine.

To know about the distribution of enzymes within the microbial world represents a valuable information in developing and improving a technical enzyme because this offers the chance to obtain enzyme variants with seperate biochemical properties. Oligo-1,6-glucosidases for instance, from strains of five various *Bacillus* species showed an increase in thermostability from mesophile, facultative thermophile, obligate thermophil, up to extreme thermophile, which corresponds with the proline content of the enzymes [63], although they are homologous proteins with respect to their catalytic and molecular properties. These enzyme variants also show various stabilities against organic solvents or denaturing concentrations of salt, and various ranges of pH-stability.

Hydroxyisocaproate dehydrogenases (HicDHs) are enzymes that reduce 2-keto acids with branched-chain or aromatic side chains enantioselectively to the corresponding hydroxy acids. They occur in strains of various species of *Lactobacillaceae* and were originally found in the course of a screening for enantioselective 2-hydroxy acid dehydrogenases [11–13]. For lactate dehydrogenases exhibit a limited substrate specificity, these alternative enzymes with their broad substrate acceptance are important for preparative purposes. The physiological significance of HicDHs are not yet clear, but such an activity could be demonstrated in several *Lactobacillus* strains. The biochemical characterization of HicDHs from various strains of *Lactobacillaceae* reveal separate enantioselectivities and kinetic properties. Most of the enzymes yield D-hydroxy acids exept the one from *Lactobacillus confusus* which is L-specific. The selectivities for keto acids are quite different, e.g. the enzyme from *Lactobacillus casei* DSM 20 008 reveals a high affinity (low K_m-value) for ketoisocaproate compared to the structurally related compounds ketomethylvalerate and ketoisovalerate. Significantly, the enzyme from this strain is the only one which shows such a high selectivity for ketoisocaproate, whereas another strain of this species, *L. casei* DSM 20 244 accepts the three keto acids with almost identical affinity.

These few examples demonstrate that at least some optimization of an enzyme, designed as a technical catalyst, can be provided by enzyme variants obtained by repeated screening steps. Basing upon biochemical and physiological data from preliminary enzyme characterization studies such improvements may be obtained by changing the screening procedure. An example for the improvement of an (R)-specific alcohol dehydrogenase is given below. Variants of this alcohol dehydrogenase with a significantly higher thermostability could be detected by an antibody-supported screening method.

Site-directed mutagenesis is a new and rapidly developing approach for the rational modification of an enzyme. This method requires exact information about the amino acid sequence, the three-dimensional structure and the active site of the enzyme. An example for a successful site-directed mutagenesis is represented by L-lactate dehydrogenase, which could be converted into an L-malate dehydrogenase by replacing Gln-102 to Arg [64]. The substrate specificity of lactate dehydrogenase from *Bacillus stearothermophilus* was extended by site-directed mutagenesis as well [65]. In some cases, the coenzyme specificity could be altered. Naturally, most of the NAD(P)-dependent dehydrogenases are highly selective for only one of the coenzymes. Some successful approaches were reported to change or extend this coenzyme specificity: The NADP-dependent glutathione reductase was altered into a NAD-specific enzyme by site-directed mutagenesis [65], and glyceraldehyde 3-phosphate dehydrogenase which depends on NAD was changed by two mutations into an enzyme accepting NADP, too [66].

Formate dehydrogenase is an important means for the regeneration of NADH. All formate dehydrogenases described so far in the literature are highly specific for NAD. Only very recently, the coenzyme specificity of the formate dehydrogenase from a *Pseudomonas* strain was altered to accept both NAD and NADP [67–69]. The preferred substrates of wild-type aspartate aminotransferase are the anionic amino acids L-aspartate and L-glutamate. This specificity was changed by the replacement of the active-site Arg by Asp, which generates an enzyme accepting the cationic amino acids L-lysine and L-arginine [70]. In the active site of yeast alcohol dehydrogenase, two amino acids with bulky side chains, Trp-93 and Thr-48 were replaced by Phe and Ser, respectively. This facilitates the oxidation of long-chain alcohols such as propanol, butanol, pentanol, hexanol, heptanol, octanol and cinnamyl alcohol [71].

Besides these rather complex coenzyme-dependent enzymes, the nonecoenzyme requiring protease subtilisin is the most extensively mutated enzyme. The substrate specificity of the enzyme as well as its dependence on pH and its stability were altered by site-directed mutagenesis [72–78]. As the knowledge about exact details of the structure and active site of the enzyme is essential for the application of this method, progress in this field is difficult to achieve. Site-directed mutagenesis as a means of catalyst improvements will be used only after extensive application of conventional optimization procedures.

2.2 Biochemical Properties of Alcohol Dehydrogenases

2.2.1 Classification of NAD(P)-Dependent Alcohol Dehydrogenases

Alcohol dehydrogenases can be subdivided with respect to various criteria, two of which are in particular relevant for the application of these enzymes: first structural and protein chemical data, especially subunit size and occurrence of metal ions, and second the stereochemical course of the catalyzed reaction and the consequential chirality of the formed alcohol.

Alcohol dehydrogenases are in generally subdivided into three groups [79], the medium-chain, zinc-containing ADHs, represented by horse liver ADH, short-chain ADHs without any metal ion, represented by the *Drosophila* ADH and the "iron-activated" long-chain ADHs with the ADH II from *Zymomonas mobilis* as the typical enzyme of this group. A recent review concerned with the molecular characterization of microbial ADHs is given by Reid and Fewson [80]. Originally, the classical yeast and liver ADHs had been termed as the long-chain ADHs in contrast to the short-chain enzymes, but were renamed as the medium-chain family [81, 82] after the still-longer ADHs had been discovered. Table 1 summarizes characteristic biochemical and microbiological data of enzymes of these groups.

Table 1. Biochemical and microbiological characterization of alcohol dehydrogenases (Classification according to [79])

Parameter	Short-chain ADH	Medium-chain ADH	Long-chain ADH, iron-activated
Amino acids per subunit	approx. 250	approx. 350	approx. 385
Metal requirements	no requirements	Zn^{2+}	Fe^{2+}
Occurrence	prokaryotes *Drosophila* ssp. mammalian tissues	pro- and eukaryotes microbes, plants, mammalian tissues	prokaryotes *Saccharomyces cerevisiae* (ADH IV)
Enzyme (Ref.) (Organism)	Sorbitol dehydrogenase [83] (*Rhodobacter sphaeroides*)	Benzyl-ADH [84] (*Acinetobacter calcoacticus*)	Glycerol dehydrogenase [85] (*Bacillus stearothermophilus*)
	LKADH [86] (*Lactobacillus kefir*)[a]	TBADH [87, 88] (*Thermoanaerobium brockii*)	Propandiol dehydrogenase [89] (*E. coli*)
	LBADH [90] (*Lactobacillus brevis*)[a]	ADH (Isoenzymes I-III) [91] (*Saccharomyces cerevisiae*)	
		HLADH [92] (horse liver)	
		Sorbitol dehydrogenase [93] (sheep liver)	
		sec.-ADH [94] (*Rhodococcus erythropolis*)	

[a] According to their amino acid content and structural relationship (see below), these enzymes are short-chain dehydrogenases, although they contain Zn^{2+} and require Mg^{2+} for their activity.

The medium-chain ADHs are defined basing upon the classical liver alcohol dehydrogenase (HLADH), which is now known to represent a large family of enzymes [82, 95], which contain frequently but not always [96] zinc at the active site. Medium-chain ADHs exist in dimeric or tetrameric forms, and depend on NAD or NADP, respectively. Alignment of amino acid sequences revealed that proteins other than ADHs are related to these enzymes, for example, threonine dehydrogenase from *E. coli* [97], human sorbitol dehydrogenase [98], glucose dehydrogenase from *Thermoplasma acidophilum* [99], or reductases such as rat [100] or yeast enoyl reductase [101] and *E. coli* quinone oxidoreductase are related to this kind of ADHs. An alignment of 106 proteins of this large enzyme family [82] revealed that only three residues are strictly conserved, which are all glycines. These residues are located in both domains and thus do not only constitute the coenzyme binding motif.

The tertiary structure of HLADH, the representing enzyme of this group, was refined to 2.4 Å resolution with X-ray diffraction [102–105], and the three-dimensional modeling of other medium-chain ADHs is based upon this structure [106, 107]. The tertiary structure of yeast ADH was found to be similar to that of HLADH although the primary structure of YADH is quite different.

The second large subgroup consists of the short-chain alcohol dehydrogenases, which have been attracting new attention in the last years because many enzymes were found to belong to this family. It was firstly defined [108, 109] with regard to the structures of *Drosophila* ADH [110, 111]], ribitol dehydrogenase from *Enterobacter aerogenes* [112, 113], and glucose dehydrogenase from *Bacillus megaterium* [114, 115]. Meanwhile more than 20 pro- and eukaryotic proteins are described to belong to this family [80]. Dimeric as well as tetrameric proteins were found, but not all these enzymes are actually alcohol dehydrogenases. Hydroxysteroid dehydrogenases, polyol dehydrogenases and proteins the biochemical function of which is not known, belong to this group, too. The enzymes may depend on NAD as well as on NADP. However related with respect to their subunit size and coenzyme binding motif, all these enzymes are very dissimilar regarding the residue identities of 15–35% only. Proteins which belong to the short-chain ADHs are not as well characterized as the medium-chain dehydrogenases. The secondary structure of the *Drosophila* ADH was predicted by Thatcher and Sawyer [110] who suggested an alternating βαβ region (a nucleotide-binding domain) at the N-terminus of the protein. This assumed function of the N-terminal part of these enzymes was supported by a conserved glycine-rich domain (GXXXGXG) found in most of the short-chain dehydrogenases.

The recent elucidation of a large number of short-chain ADH sequences permits a detailed characterization of this enzyme group. As many as six domains are found to be significantly conserved. A domain located at the N-terminus with a length of approximately 30 amino acids is generally assumed to be the coenzyme binding site. A second conserved domain is a hydrophobic region comprising 10 or 11 residues, respectively. A further conserved domain consists of 18 amino acids are probably involved in the dimerization of enzyme subunits or facilitate the hydride transfer. It starts with a totally conserved tyrosine and ends with a highly conserved acidic amino acid.

The third subgroup of alcohol dehydrogenases consists of "iron-activated" enzymes. The first enzyme detected to belong to this group was the ADH II from *Zymomonas mobilis* [116] followed by the observation that the ADH IV from *Saccharomyces cerevisiae* shows more than 50% identity to this bacterial ADH [117]. No data about the secondary or tertiary structures of the enzymes in this subgroup are available currently. A prediction based upon the Chou and Fasman analysis [118] indicates that these enzymes are rich in α-helices.

For a classification of the ADHs according to the stereochemistry of the reduction one has to consider that the reduced coenzymes NADH and NADPH possess two diastereotopic hydrogens, pro-*R* (A-specific dehydrogenases) and pro-*S* (B-specific enzymes) and that the substrate ketone can be attacked by two sides. It is possible to determine which hydrogen is transferred by the enzyme using [4-^{2}H]-labelled NAD(P)H for the reduction reaction or labelled alcohol and NAD(P) [119, 120]. This hydrogen can attack the carbonyl group from the *re*- or the *si*-side, resulting in an (*S*)- or (*R*)-alcohol. Table 2 shows the various stereochemical courses of the hydrogen transfer and representative enzymes for the different cases. Most alcohol dehydrogenases (sometimes termed as A-enzymes) transfer the pro-*R* hydrogen to the *re*-face of substrate ketone forming (*S*)-alcohol, a process characterized by Prelog's rule [37]. Table 2 also illustrates that neither the kind of hydrogen transfer nor the stereochemistry of the carbonyl reduction is correlated to a preference of the kind of coenzyme (NAD or NADP) involved.

Table 2. Stereochemistry of the hydrid transfer catalyzed by alcohol dehydrogenases

Parameter	Enzyme group			
	E$_1$	E$_2$	E$_3$	E$_4$
Transferred hydrogen	pro-*R*	pro-*S*	pro-*R*	pro-*S*
Face of the attacked carbonyl	*si*-face	*si*-face	*re*-face	*re*-face
Chirality of the formed alcohol	(*R*)-alcohol	(*R*)-alcohol	(*S*)-alcohol	(*S*)-alcohol
Alcohol dehydrogenases (Coenzyme)	*Lactobacillus kefir* (NADP)	*Mucor javanicus* (NAD)	yeast (NAD)	not known
	Pseudomonas sp. (NAD)		horse liver (NAD)	
			Thermoanaerobium brockii (NADP)	

2.2.2 Characterization of Available Alcohol Dehydrogenases

Many NAD(P)-dependent alcohol dehydrogenases from microorganisms, plant and mammalian cells have been purified and biochemically characterized so far. Most of them may be classified into three subgroups, and they represent a large database to give information about phylogenetic and evolutionary relationships and to allow comparative biochemical studies on the function of conserved sequences and structure-function relationships. Only few of these enzymes are important for preparative applications, particularly those that are commercially available or easy to prepare. Among the commercially available ADHs, there are some like the hydroxysteroid dehydrogenases or glycerol dehydrogenase that show a very narrow substrate specificity. Nevertheless, the preparation of some chiral alcohols in the gram-scale has already demonstrated the applicability of these enzymes for a limited number of substrates. Besides these specific enzymes, there are three alcohol dehydrogenases, isolated from horse liver (HLADH), yeast (YADH) and *Thermoanaerobium brockii* (TBADH) that are important because they accept a broad spectrum of substrates or are available at a very low price (Table 3).

Horse Liver ADH

Horse liver ADH is very useful for preparative applications. It is one of the oxidoreductases, which have been studied most exceedingly. It is a dimer consisting of nearly identical subunits. The subunits are designated as E (for

Table 3. Characterization of commercially available alcohol dehydrogenases which accept a rather broad spectrum of substrates

	Yeast ADH	Horse liver ADH	T. brockii ADH
Classification	medium-chain dehydrogenase	medium-chain dehydrogenase	medium-chain dehydrogenase
Required coenzyme	NADH	NADH	NADPH
Preferred substrates	aldehydes	aldehydes, cycl. ketones	ketones
Specific activity (U/mg)	300[a]	1–2[a]	30–90[b]
Enzyme costs ($/1000 Unit) (Sigma catalogue 1996)	1.0	530	870
Stability	sensitive to O_2	stable	stable
Limitations	low stability		low enantio-selectivity for some substrates; NADP-dependence; costs

[a] Oxidation of 1 μmol ethanol per min
[b] Oxidation of 1 μmol isopropanol per min

ethanol active) with 374 amino acids, and S (for steroid active), which differs from the E unit in only a six amino acid residue [103, 105]. There are three possible species of the enzyme, EE, SS and ES, however the EE form is dominating in commercially available highly purified products. Each subunit contains two zinc atoms. The X-ray structure was obtained with the EE isozyme.

HLADH converts a wide range of substrates. For the predicition of the stereoselectivity of reduction reactions, originally Prelog's diamond lattice model was applied, which is based upon the characteristic properties of the ADH of *Curvularia falcata* [37]. This model describes the stereospecificity of HLADH catalyzed reductions of simple acyclic substrates such as aldehydes. Later on, for more complex acyclic and cyclic substrates, a cubic-space model of the active site was developed [38, 121]. Other models are based upon symmetric properties [122–125] or upon a refined diamond lattice model [126–129].

Yeast ADH

Yeast ADH is a tetramer containing varying amounts of zinc. Although the amino acid sequence is quite different from HLADH, tertiary structures seem to be similiar in both enzymes [103, 105, 130, 131]. Yeast ADH has a very narrow substrate specificity and usually accepts aldehydes and methyl ketones, only [132, 133]. Compared to the other commercially available ADHs YADH is a cheap enzyme; but it is quite unstable. Thus, YADH is only of limited use for the preparation of chiral alcohols. Because of its low cost, YADH has been studied thoroughly regarding its potential for the regeneration of the reduced coenzymes [134–137], but ethanol being the regeneration substrate as well as acetaldehyde being the product deactivate YADH at very low concentrations.

Thermoanaerobium brockii ADH

TBADH is an NADP-dependent dehydrogenase which oxidizes preferably secondary alcohols. It was isolated from a thermophilic microorganism and exhibits a remarkable thermostability [138–140]. The enzyme remains stable up to 65°C. Since neither YADH nor HLADH are able to convert open-chain secondary alcohols, TBADH fills this gap: its reactivity is highest against secondary alcohols, and decreasing against linear and cyclic ketones and being low against primary alcohols. Because of this substrate spectrum, the stereospecific reduction of ketones can be coupled with the TBADH-catalyzed oxidation of isopropanol to regenerate NADPH. The enantiomeric excess of the product alcohols obtained by reduction with TBADH decreases if small substrates like methylethylketone are converted. Table 3 summarizes basical properties of the three commercially available alcohol dehydrogenases that are important for preparative applications.

Glycerol Dehydrogenases

Purified glycerol dehydrogenases are commercially available from several microbial sources, for instance *Bacillus megaterium* (NAD-dependent; 20 U/mg (glycerol oxidation)), *Cellulomonas* sp. (NAD; 50–125 U/mg), *Enterobacter aerogenes* (NAD; 25 U/mg) or *Aspergillus niger* (NADP; 10–20 U/mg), respectively. *In vivo*, these enzymes catalyze the interconversion of glycerol and dihydroxyacetone, *in vitro* they have been used for the enantioselective reduction of achiral 2-hydroxy ketones to chiral 1,2-diol, the kinetic resolution by reduction of racemic 2-hydroxy ketones and the stereoselective oxidation of meso-1,2-diols to 2-hydroxy ketones [141]. The most important application of glycerol dehydrogenases is the enzyme-catalyzed reduction of prochiral hydroxy ketones, whereas the oxidation reactions are limited by a severe product inhibition. Reduction of 1-hydroxy-2-propanone or 1-hydroxy-2-butanone to the corresponding (R)-diols are described with ee-values of 95–97% on a gram-scale [141].

Since glycerol dehydrogenase contains autoxidizable thiol groups, it is nessecary to perform these reaction in an inert atmosphere; antioxidants such as dithiothreitol or mercaptoethanol deactivate the enzyme at higher concentrations and should be avoided in preparative applications.

Hydroxysteroid Dehydrogenases

Several hydroxysteroid dehydrogenases (HSDHs) are available with various regioselectivities. They are named by their ability to oxidize or reduce various hydroxylated steroid structures, for instance 3α-hydroxysteroid dehydrogenase oxidizes androsterone at the 3α position. It can be isolated from *Pseudomonas testosteroni*. Commercially available samples with a specific activity of 15–50 U/mg are free of 3β-hydroxysteroid dehydrogenases. Other commercially available enzymes belonging to this group are 7α-HSDH from *Escherichia coli* or *Pseudomonas* sp., 12α-HSDH from *Bacillus sphaericus*, and 3α,20β-HSDH (cortison reductase) from *Streptomyces hydrogenans*. Keto groups of steroid and bile acid molecules can be reduced on a preparative scale regio- and stereospecifically using hydroxysteroid dehydrogenases [142–144]. The transformations are complete and virtually pure products are obtained. Reactions with neutral steroids, which are poorly soluble in aqueous buffers, are carried out in water-organic solvent two-phase systems [143]. Some of these enzymes catalyze transformations of non-steroid ketones, too. In particular, 3α-HSDH appears to be useful in oxidizing aromatic *trans*-diols with high enantioselectivity. All the commercially available HSDHs are rather expensive.

Diketone and Other Ketone Reductases from Microbial Sources

Diacetyl reductase (acetoin dehydrogenase) isolated from *Lactobacillus kefir* converts the prochiral diacetyl into optically pure (+)-acetoin (ee > 94%) [145].

This NAD-dependent enzyme was purified up to a specific activity of 1060 U/mg (diacetyl as substrate). The enzyme is stable at 57°C for 10 min, the temperature optimum is at 70°C. Besides diacetyl several other diketones were reduced.

From *Candida parapsilosis* an NADP-dependent carbonyl reductase was purified which can reduce a variety of diketones such as indole-2,3-diones and analogues, or dihydro-4,4-dimethyl-2,3-furandiones [146]. The reduction of the latter compound gives (R)-(-)-dihydro-3-hydroxy-4,4-dimethyl-2(3H)-furanone, a key intermediate in the synthesis of D-pantothenic acid. The reduction of 2-(6-carbomethoxyhexyl)cyclopentane-1,3,4-trione results in important intermediates of (-)-prostaglandin E_1 and (-)-prostaglandin E_2. An NAD-dependent carbonyl reductase from *Candida parapsilosis* converts a broad variety of carbonyl compounds resulting in (S)-hydroxy compounds [147, 148]. A carbonyl reductase from *Mucor ambiguus* is strictly specific for conjugated polyketone compounds only [149]. Similar enzymes seem to be widely distributed among microorganisms [146, 150–152]. Out of all these enzymes, only diacetyl reductase from *Lactobacillus kefir* is commercially available.

ADH from *Sulfolobus solfataricus*

An NAD-dependent alcohol dehydrogenase was isolated from the extreme thermophilic archaebacterium *Sulfolobus solfataricus* [59]. The enzyme has a broad substrate specificity that includes linear and branched primary alcohols, linear and cyclic secondary alcohols and linear and cyclic ketones. The enzyme is a dimer with a molecular weight of 37.6 kDa, it contains four zinc atoms per dimer. The gene encoding this ADH has been isolated and the primary structure, determined by peptide and gene analysis consists of 347 amino acids [60]. Due to these structural features it belongs to the medium-chain ADH family. The enzyme activity increases with temperature up to 95°C. At 60°C and 70°C the half-life was 20 h and 5 h, respectively. The specific activity of the highly purified enzyme is in the range of 3.9 U/mg, measured with benzyl alcohol at 65°C. *Sulfolobus* ADH is not available commercially.

Enoate Reductases

Enoate reductase [153, 154], which occurs in strains of *Clostridium* or *Proteus*, and 2-oxo-acid reductase [155] from *Proteus vulgaris* or *P. mirabilis* catalyzes the stereospecific reduction of substrates performed directly by reduced methyl-viologen. No nicotinamide coenzyme is required. Methylviologen is regenerable electrochemically. Examples of the reduction of enoates, ketones, and 2-oxo acids are given in [155].

2.3 New Alcohol Dehydrogenases from Strains of Rhodococcus and Lactobacillus

None of the commercially available or described alcohol dehydrogenases is able to reduce ketones with bulky side chains such as acetophenone or pinacoline. TBADH when tested with acetophenone [156] was found to be inactive. As optically active phenylethanol and its derivatives are important chiral building blocks, we carried out a limited screening among bacteria and yeasts from a culture collection, in order to isolate NAD(P)-dependent dehydrogenases. This screening resulted in several microorganisms which showed reduction activity for acetophenone or p-Cl-acetophenone in the crude extract [88, 157, 158]. Preliminary results with the most active strains *Rhodococcus erythropolis* DSM 43 297 and *Lactobacillus kefir* DSM 20 587 revealed that these enzymes converted acetophenone and its p-Cl-derivative with an enantiomeric excess of 100%, but yielded in opposite isomers; the enzyme from *Rhodococcus erythropolis* gives the (*S*)-phenylethanols, whereas the *L. kefir* results in the corresponding (*R*)-alcohols [157]. The coenzyme specificity was found to be different, too; the (*S*)-ADH depends on NAD, whereas the (*R*)-ADH accepts NADP only. Both enzymes were purified and characterized and they both proved to be suitable for preparative applications. The characterization of the (*R*)-ADH from *L. kefir* also included microbiological and physiological studies and the reactions with polyclonal antibodies, which led to the detection of a further (*R*)-ADH producer, *Lactobacillus brevis*, which possesses a dehydrogenase with a significantly higher stability.

2.3.1 (S)-Alcohol Dehydrogenase from Rhodococcus erythropolis DSM 43 297

S-ADH from *Rhodococcus erythropolis* DSM 43 297 is an important catalyst for the preparation of enantiomerically pure alcohols. It accepts ketones with bulky side chains such as acetophenone, ring-halogenated acetophenones, or pinacoline. An enzyme sample which is already sufficient for technical applications can be obtained by a single chromatographic purification step with an anionic exchanger. The resulting crude enzyme preparation has a specific activity of about 30 U/mg, tested with p-Cl-acetophenone. After four chromatographic purification steps and application of a preparative electrophoresis, the enzyme preparation is pure when tested electrophoretically (Table 4). The specific activity of this sample is 1400 U/mg. Furthermore, the enzyme proved to be quite stable. In a continuous production process, no significant loss of activity could be observed within one week [159]. Some substrates that can be reduced by this enzyme are given in Table 5. (*S*)-ADH from *Rhodococcus erythropolis* converts not only acetophenone and its derivatives and β-ketoesters, but also α-ketoesters (Table 6). Methyl- and ethylpyruvate are rather very good substrates for this enzyme, related to p-Cl-acetophenone they are converted with 13- or 8-fold reaction velocities, respectively. Optically pure α-hydroxyesters are

Table 4. Purification of (S)-ADH from *Rhodococcus erythropolis* DSM 43 297 (1 U = reduction of 1 μmol p-Cl-acetophenone per min; HIC = hydrophobic interaction chromatography; Prep-Cell = preparative gel electrophoresis)

Purification step	Total units (U)	Total protein (mg)	Specific activity (U/mg)	Yield (%)
Crude extract	254	11.5	22	100
Q-Sepharose FF	271	8.3	32.5	100
Phenyl-650 C (HIC)	182	2.35	77	72
Butyl-Sepharose 4FF	142	1.14	124	56
Prep-Cell[a]	27	0.004	1416	21

[a] applying 50% of the material from the Butyl-Sepharose chromatography

Table 5. Substrate specificity of the (S)-ADH from *Rhodococcus erythropolis* DSM 43 297

Substrate	Concentration [mM]	Relative Activity
Acetophenone	3	27
p-Cl-Acetophenone	3	100
o-Cl-Acetophenone	3	12
m-Cl-Acetophenone	3	294
p-Br-Acetophenone	3	196
p-F-Acetophenone	3	41
p-Methyl-Acetophenone	3	122
p-Methoxy-Acetophenone	3	37
1-Phenyl-propan-2-one	10	12
4-Phenyl-butan-2-one	12	80
6-Methyl-hept-5-en-2-one	12	73

Table 6. Kinetic properties of (S)-ADH from *Rhodococcus erythropolis* for the reduction of α-ketoesters

Substrate	Activity at 1 mM substrate conc. (U/mg)	K_m value (mM/L)	v_{max} (%)[a]
p-Cl-Acetophenone	13.0	—	100
Methylpyruvate	7.8	17.6	1300
Ethylpyruvate	18.6	2.1	823
Ethylbromopyruvate	6.3	—	—
Ethyl-3-methyl-2-oxobutyrate	4.7	—	—
Phenylglyoxalic acid ethylester	3.4	—	—
Phenylglyoxalic acid methylester	3.6	—	—
2-Oxo-4-phenylbutyric acid ethylester	5.0	—	—

[a] in relation to the v_{max} value (13 U/mg) obtained with p-Cl-Acetophenone (measured at pH 6.00)

important chiral building blocks or starting compounds of chiral α-hydroxy acids, which can be released from the esters by cleavage of the ester bond. All hydroxy compounds produced so far have an enantiomeric excess of more than 99%. (S)-ADH from *Rhodococcus erythropolis* has a molecular weight of

160 kDa and consists of four identical subunits. With regard to its molecular properties, it belongs to the group of medium-chain ADHs.

2.3.2 (R)-Alcohol Dehydrogenase from Lactobacillus Strains

(R)-Alcohol dehydrogenase from L. kefir was found to reduce acetophenone with a high activity even when used as crude extract. The activity with unpurified cell-free samples was in the range of up to 10 U/mg (reduction of acetophenone). For preparative applications, a technical enzyme sample can simply be reached by only one chromatographic purification step [88]. No disturbing or interfering activities were observed, thus the purification obviously is a suitable means to remove low-molecular weight compounds interfering with the substrate and to standardize and concentrate the enzyme. After chromatography on an anionic exchanger (Mono Q), samples with specific activities of 30 U/mg and more could be obtained (80–90% yield; 5–20-fold enrichment, depending on the starting material). These preparations were suited for first studies about the applicability of the enzyme [88, 160].

2.3.2.1 Purification and Biochemical Characterization of (R)-ADH from Lactobacillus kefir

A homogeneous enzyme sample was obtained after four chromatographic purification steps (Table 7) including affinity chromatography on 2',5'-AMP-sepharose. In spite of optimization of the purification procedure, severe losses of activity could not be avoided; the yield of the activity of the pure enzyme was in the range of 0.1–3%. The specific activity of the homogeneous protein was in the range of 180 U/mg, when measured with acetophenone as the substrate. The decrease of activity was due to an increasing instability of enzyme in the course of the enrichment. This instability was not a result of the cation dependence of the enzyme. During the first efforts to purify the enzyme it was observed that the activity strictly depends on Mg^{2+} or Mn^{2+}, the addition of 1mM of Mg^{2+} prevents completely the loss of activity.

Table 7. Purification of alcohol dehydrogenase from *Lactobacillus kefir*. (Phenyl- and octylsepharoses are materials for hydrophobic interaction chromatography; Mono Q is an anionic exchanger.)

Purification step	Activity [U/ml]	Specific activity [U/mg]	Yield [%]
Crude extract	32.9	6.3	100
Phenylsepharose	9.7	17.8	23
Octylsepharose	1.7	28.0	10
Mono Q	6.5	97.0	1.7
2',5'-AMP-Sepharose	10.7	149.0	1.0

The dependence of metal ions is an important feature for the classification of ADHs. Thus, the existence of a weakly bound ion in the *L. kefir* ADH was unexpected and needed assurance by the application of selectively chelating inhibitors. During purification of the crude extract by ion exchangers it was observed, that the activity was almost completely lost. Addition of several cations to such partially inactivated samples resulted in a significant reactivation (130%) by Mg^{2+}-ions. Studies with cation chelating compounds revealed that the ADH from *L. kefir* contains a weakly bound cation. Addition of EDTA deactivates the enzyme completely, whereas chelators for transition metals such as 2,2-dipyridine or 1,10-phenanthroline show no influence on the activity at all. According to the classification of biologically important metals [161], the s-block elements sodium, potassium, magnesium, and calcium interact only weakly with ligands other than oxygen, whereas the transition metals, with partly filled *d*-orbitals, and zinc, bind nitrogen ligands much more strongly. The results with the ADH from *L. kefir* support the assumption that Mg^{2+} is essential for its activity. Purification of the ADH is possible when 1mM of Mg^{2+} is added. In spite of the instability of the ADH from *L. kefir* it was possible to obtain pure enzyme protein.

The preparation of chiral alcohols can be carried out very simply because the regeneration of NADPH is possible by the addition of isopropanol. Unpurified crude extract samples of the ADH from *L. kefir* were found to be a useful catalyst for the synthesis of (*R*)-alcohols [160]; some examples for the preparation of some chiral alcohols using this enzyme are given in Table 8. Though this ADH becomes unstable to such a degree during the purification process, enough material of the pure enzyme could be prepared to produce polyclonal antibodies and to screen for related (*R*)-specific enzymes.

2.3.2.2 Antibody-Supported Screening for (*R*)-Specific Alcohol Dehydrogenases

Several screening procedures among microorganisms were carried out to find enzymes which are capable to reduce acetophenone. The detection of an (*R*)-specific alcohol dehydrogenase in *L. kefir* was the first success of these efforts. Polyclonal antibodies permit to look for the distribution of a given protein among other organisms and to screen for related enzymes. Assays with the commercially available alcohol dehydrogenases from yeast, horse liver and *Thermoanaerobium brockii* indeed, gave no reaction with the antibody; but screening among the genus *Lactobacillus* revealed that each strain of the subgroup *Betabacterium* gave positive results whereas strains of the other subgroups were found to be inactive (Table 9). All the antibody-positive *Betabacterium* strains were assayed with regard to their enzyme activities with acetophenone as the substrate. Only two positive strains were detected, *L. kefir* and *L. brevis*. (Table 10). These results demonstrate that probably each strain of the subgroup of *Betabacterium* possesses a characteristic antibody-positive dehydrogenase, whose biological function is not yet known. The substrate

Table 8. Preparation of chiral alcohols by enzyme-catalyzed reduction of the corresponding ketones with ADH from *Lactobacillus kefir*. The production of phenylethanol with formate and formate dehydrogenase (FDH) for coenzyme regeneration was carried out continuously in an enzyme-membrane-reactor

Alcohol	Method for coenzyme regeneration	Ketone concentration [mM]	Yield [%]	Enantiomeric excess [%]	Ref.
(S)-1-Phenyl-2,2,2-trifluoroethanol	isopropanol	150	71	> 99	[160]
(R)-1-(2-Pyridyl)-ethanol	isopropanol	150	60	> 97	[160]
(R)-1-(2-Furanyl)-ethanol	isopropanol	150	65	95	[160]
(R)-6-Methyl-5-hepten-2-ol	isopropanol	150	58	> 99	[160]
(R)-5-Chloro-2-pentanol	isopropanol	150	52	> 97	[160]
(R)-1-Cyclopropyl-1-ethanol	isopropanol	150	46	> 97	[160]
(R)-5-Norbornen-2-ol	sopropanol	150	39	> 97	[160]
(R)-1-(Trimethylsilyl)-1-butyn-3-ol	isopropanol	150	25	94	[160]
Methyl-4-hydroxy-1-(trimethylsilyl)-5-hexynoate	isopropanol	150	15	97	[160]
(R)-p-Cl-Phenylethanol	isopropanol	10	100	100	[94]
(R)-Phenylethanol	isopropanol	5	100	100	[94]
(R)-Phenylethanol	formate/FDH	10	> 90	100	[67]
(R)-4-Phenyl-2-butanol	formate/rhodium-complex	33	81	96	[162]

Table 9. Taxonomic classification of *Lactobacillus* strains according to Bergey's Manual [163] and their reaction with the *anti-L. kefir*-ADH antibody. *Boldface-typed* strains were tested, *underlined strains* gave a positive reaction with the antibody (*L.* = *Lactobacillus*)

Thermobacterium	Streptobacterium	Betabacterium	
		Group A	Group B
L. acidophilus	*L. casei*	*L. kefir*	*L. hilgardii*
L. helveticus	*L. plantarum*	*L. brevis*	*L. fructivorans*
L. bulgaricus	*L. alimentarius*	*L. cellobiosus*	*L. trichodes*
L. delbrueckii	*L. curvatus*	*L. fermentum*	*L. desidiosus*
L. salivarius	*L. coryneformis*	*L. buchneri*	*L. heterohiochi*
L. leichmanii	*L. farciminis*	*L. viridescens*	
L. lactis	*L. homohiochi*	*L. confusus*	
L. jensenii	*L. xylosus*		

Table 10. Enzyme activity of strains of the betabacterium subgroup A, tested with acetophenone and NADPH

Strain	Activity [U/mg]
L. kefir	87.0
L. brevis	93.0
L. cellobiosus	0.9
L. fermentum	0.2
L. viridescens	0.2
L. confusus	0.3
L. buchneri	0.8

specificity of these dehydrogenases however vary in the various strains of the subgroup, and two of them, L. kefir and L. brevis, are able to reduce ketones with bulky side chains such as acetophenone. In order to get an impression about the biological function of this dehydrogenase, the physiological characterization of this subgroup is summarized shortly.

The genus Lactobacillus was devided by Orla-Jensen [164, 165] into three main subgroups, Thermobacterium, Streptobacterium and Betabacterium basing upon the optimal growth temperatures and fermentation end products. This division has been confirmed by additional physiological tests [166, 167]. Strains of the subgroup Betabacterium metabolize glucose in a heterofermentative way to produce lactic acid, CO_2, ethanol and/or acetic acid. The production of CO_2 is a means in practice to distinguish betabacteria from the homofermentative lactobacilli. Further properties of betabacteria are the thiamine requirement for growth, the production of mannitol as an end product of the fructose fermentation and the absence of fructose-1,6-bisphosphate aldolase and triosephosphate isomerase [168, 169]. As these enzymes are lacking, which are essential for the glucose degradation via the fructose-1,6-bisphosphate pathway, glucose is metabolized exclusively by the pentose-phosphate pathway. Subgroup A of the betabacteria consists of fermentatively active species, whereas strains of the subgroup B are inert to most carbohydrates, but grow at pH values as low as 3.2 and on 15% ethanol [163].

2.3.2.3 Induction of the (R)-Alcohol Dehydrogenase

Growth studies of L. kefir on various carbon sources revealed, that this organism is able to grow likewise on glucose, arabinose, ribose, and lactose, but enzyme activity was found only on glucose and the glucose containing lactose (Table 11). The induction of alcohol dehydrogenase of L. kefir by glucose could be demonstrated in the strain by growth on arabinose and different amounts of glucose (Figure 2). Growth in the presence of 0.5% of glucose was found to be optimal for the enzyme activity. In order to obtain information about the kind of activation of the enzyme, crude extracts were prepared by growth on glucose

Table 11. Growth of *Lactobacillus kefir* on various carbon sources (2%) and activity of (R)-ADH measured with acetophenone and NADPH. (Optical density was measured at 660 nm)

Carbon source	Growth (Optical density)	Enzyme activity [U/mg]
Glucose	3.4	4.9
Arabinose	3.8	0.7
Ribose	2.0	0
Lactose	3.3	2.1

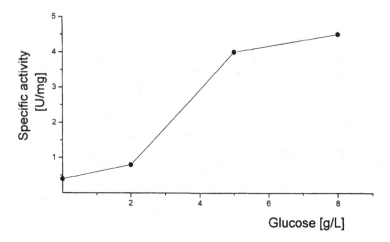

Fig. 2. Induction of (R)-alcohol dehydrogenase during growth of *Lactobacillus kefir* on increasing concentrations of glucose

or arabinose, respectively. These crude extracts were used both to demonstrate the enzymatic ability to reduce actophenone and to test the reactivity against polyclonal anti-LKADH antibodies. Within a 3 h incubation, 10 mM of aceto-phenone were converted completely to (R)-phenylethanol with the enzyme extract of cells grown on glucose, whereas no phenylethanol was found within this time with arabinose-grown cells. The antibody-assay was positive with the glucose-grown cells only, no reaction was found with the extract of arabinose-grown cells. Thus it seems evident, that the activation of the alcohol dehydro-genase by glucose occurs at the DNA level.

2.3.2.4 Purification of Alcohol Dehydrogenase from *Lactobacillus brevis*

The ADH from *Lactobacillus brevis* can be purified by exactly the same chromatographic procedures as applied for the purification of *L. kefir* ADH.

However, the enzyme from *L. brevis* could be eluted from the anionic exchanger at a significantly lower ionic strength, which suggests a different surface charge of both proteins. After application of four chromatographic steps, the enzyme is homogeneous (Table 12). Starting the purification of the *L. brevis* enzyme with the same amount of protein as for the *L. kefir* enzyme, the overall yield was about 10-fold higher due to the higher stability of the *L. brevis* alcohol dehydrogenase.

Enzyme assays with various substrate structures show that a broad range of compounds may be reduced with high activity (Table 13). Thus, the (*R*)-ADHs

Table 12. Purification of alcohol dehydrogenase from *Lactobacillus brevis*. (Phenyl- and octylsepharoses are materials for hydrophobic interaction chromatography; Mono Q is an anionic exchanger.)

Purification step	Activity [U/ml]	Specific activity [U/mg]	Yield [%]
Crude extract	89	30.0	100
Phenylsepharose	57	64.2	51
Octylsepharose	34	86	22.8
Mono Q	37	211	16.6
2′, 5′-AMP-Sepharose	64	306	9.6

Table 13. Substrate specificity of alcohol dehydrogenase from *Lactobacillus brevis*

Substrate	Relative Activity [%]
Acetophenone and -derivatives	
Acetophenone	100
4-Cl-Acetophenone	203
3-Cl-Acetophenone	147
2-Cl-Acetophenone	10
4-Ethylacetophenone	65
Propiophenone	17
Benzaldehyde	9
Benzylaceton	99
Methyl-naphthylketone	37
Aliphatic, open-chained ketones	
2,4-Pentandione	130
Hydroxyaceton	23
2-Ketoesters	
Methylpyruvate	98
Ethylpyruvate	248
3-Ketoesters	
3-Ketovaleric acid ethylester	70
3-Ketovaleric acid methylester	69
Cyclic ketones	
2-Methyl-cyclohexanone	192

Table 14. Biochemical properties of the ADHs from *L. kefir* and *L. brevis*

Parameter	*L. kefir*-ADH	*L. brevis*-ADH
Specific activity (acetophenone/NADPH)	760 U/mg	490 U/mg
Coenzyme	NADPH	NADPH
K_m (acetophenone)	0.36 mM	0.85 mM
K_m (NADPH)	0.13 mM	0.16 mM
Molecular mass	105 kDa, homotetramer	105 kDa, homotetramer
Temperature stability (30 min)	37°C	65°C
Temperature optimum	37°C	55°C
pH-optimum (reduction of acetophenone)	7.0	7.0
Dependence on cation	Mg^{2+} (Mn^{2+})	Mg^{2+} (Mn^{2+})

from strains of *Lactobacillus* are useful catalysts for the synthesis of chiral alcohols. The biochemical properties of the (*R*)-ADHs from *Lactobacillus kefir* and *brevis* and data important for their preparative applications are summarized in Table 14.

2.3.2.5 N-Terminal Amino Acid Sequences of (*R*)-Alcohol Dehydrogenases from *L. kefir* and *L. brevis*

The partial N-terminal amino acid sequences obtained by stepwise automated degradation of chromatographically pure ADHs from *L. kefir* and *L. brevis* are summarized in Fig. 3. Both sequences are given at a length of about 45 amino acids of the N-terminus as compared to the sequences of other ADHs. Obviously there is a strong homology of short-chain ADHs in this region to the (*R*)-ADHs from both *Lactobacillus* strains. All related proteins belong to the subgroup of short-chain ADHs.

Within the sequence of the first 40 amino acids of the N-terminus, which is generally regarded as the coenzyme-binding site, six amino acid differ from each other in the *L. brevis* and *L. kefir* ADHs. Three of them are responsible for differing ionic properties of this region, Asn-3 (*L. brevis*) changed into Asp (*L. kefir*), Asp-6 into Lys, and Thr-25 into Asp. The coenzyme-binding sequence G-G-T-L-G-I-G found at the positions 14–20 of *L. brevis* ADH is identical in both enzymes.

2.3.2.6 *L. brevis* ADH Gene and Protein Sequence Determination

After digestion of the homogeneous enzyme with LysC protease and separation of the peptides by HPLC, the sequences of these fragments revealed the C-terminus. The ADH gene was isolated by screening of genomic libraries of *L. brevis* with oligonucleotide probes. The encoding sequence consists of 750 base pairs. Figure 4 summarizes the complete sequence of the *L. brevis* ADH as determined by DNA and peptide sequence analysis. The primary structure consists of 250 amino acid residues.

```
                      +  *+ ++  **+   *+* + +    + +** *           +
ADH-Lb         MSNRLDGKVAIITGGTLGIGLAIATKFVEEGAKVMITGRHSDVGEK

ADH-Lk          TDRLKGKVAIVTGGTLGIGLAIADKFVEEGAKVVITGRHADVGEK

ADH-Zm       (41)VPTPMPKRLDGKVAIVTGGARGIGEAIVRLFAKHGARVVIADIDDAAGEA

7HSDH-Ec       MFNSDNLRLDGKCAIITGAGAGIGKEIAITFATAGASVVVSDINADAANH

SDH-Rs         MRLDGKTALITGSARGIGRAFAEAYVREGARVAIADINLEAARE

PGDH-Hs        MHVNGKVALVTGAAQGIGRAFAEALLLKGAKVALVDWNLEAGVQ

AMDG-Fs     TTAGVSRRPGRLAGKAAIVTGAAGGIGRATVEAYLREGASVVAMDLAPRLAAT

BDDDH-Ps       MKLKGEAVLITGGASGLGRALVDRFVAEGAKVAVLDKSAERLAE

KACPR-Ec       MNFEGKIALVTGASRGIGRAIAETLAARGAKVIGTATSENGARI

DCHDDH-Pp      MSDLSGKTIIVTGGGSGIGRATVELLVASGANVPVADINDEAGEA

3BHDH-Ct       TNRLQGKVALVTGGASGVGLEVVKLLLGEGAKVAFSDINEAAGQQ

20BHDH-Sh      MNDLSGKTVIITGGARGLGAEAARQAVAAGARVVLADVLDEEGAA
```

Fig. 3. Alignment of the *N*-terminal sequence of the ADH from *Lactobacillus brevis* and homologues proteins.
*, Identity of an amino acid in all proteins; +, identity of an amino acid in six or more proteins.
ADH-Lb, alcohol dehydrogenase from *Lactobacillus brevis* [94]; ADH-Lk, alcohol dehydrogenase from *Lactobacillus kefir* [94]; ADH-Zm, alcohol dehydrogenase from *Zea mays* [170]; 7HSDH-Ec, 7α-hydroxysteroid dehydrogenase from *Escherichia coli* [171]; SDH-Rs, sorbitol dehydrogenase from *Rhodobacter sphaeroides* [84]; PGDH-Hs, human 15-hydroxyprostaglandin dehydrogenase [172]; AMDG-Fs, *N*-acylmannosamine-1-dehydrogenase from *Flavobacterium* sp. [173]; DBDDH-Pp, *cis*-1,2-dihydrobenzene-2,2-diol dehydrogenase from *Pseudomonas putida* [174]; BDDDH-Ps, biphenyl-2,3-dihydrodiol dehydrogenase from *Pseudomonas* sp. [175]; KACPR-Ec, ketoacyl-acyl carrier protein reductase homologue from *Escherichia coli* [176]; 3BHDH-Ct, 3β-hydroxysteroid dehydrogenase from *Comomonas testosteroni* [177]; 20BHDH-Sh, 20β-hydroxysteroid dehydrogenase from *Streptomyces hydrogenans* [178]

The primary structure of the ADH from *L. brevis* contains several structures which are typical for short-chain ADHs. The N-terminus, with a length of approximately 30 amino acids, is widely regarded as the coenzyme binding site with the conserved motif G-X-X-X-G-X-G, which is G-G-T-L-G-I-G for *Lactobacillus brevis*. A second conserved domain found in the *L. brevis*-ADH sequence is a hydrophobic region comprising 10 or 11 residues, respectively. It contains two highly conserved glycines (G_{82} and G_{92}), separated by nine amino acids. Such structures seem to be located inside the protein and determine the conformation of the enzyme.

A third conserved domain identified in the *L. brevis* ADH begins with a totally conserved tyrosine (Y_{156}). It consists of 18 amino acids ending with a highly conserved acidic amino acid, which is asp in this case. This domain also contains the highly conserved lysine (K_{160}), forming together with tyr-156 a motif previously found in several studies [172, 179, 180]. These amino acid residues are probably involved in the dimerization of enzyme subunits [115] or facilitate the hydride transfer from the alcohol to the cofactor. Crystallographic

ATG TCT AAC CGT TTG GAT GGT AAG GTA GCA ATC ATT ACA GGT GGT ACG
 M   S   N   R   L   D   G   K   V   A   I   I   T   G   G   T
TTG GGT ATC GGT TTA GCT ATC GCC ACG AAG TTC GTT GAA GAA GGG
 L   G   I   G   L   A   I   A   T   K   F   V   E   E   G
GCT AAG GTC ATG ATT ACC GGC CGG CAC AGC GAT GTT GGT GAA AAA
 A   K   V   M   I   T   G   R   H   S   D   V   G   E   K
GCA GCT AAG AGT GTC GGC ACT CCT GAT CAG ATT CAA TTT TTC CAA CAT
 A   A   K   S   V   G   T   P   D   Q   I   Q   F   F   Q   H
GAT TCT TCC GAT GAA GAC GGC TGG ACG AAA TTA TTC GAT GCA ACG GAA
 D   S   S   D   E   D   G   W   T   K   L   F   D   A   T   E
AAA GCC TTT GGC CCA GTT TCT ACA TTA GTT AAT AAC GCT GGG ATC
 K   A   F   G   P   V   S   T   L   V   N   N   A   G   I
GCG GTT AAC AAG AGT GTC GAA GAA ACC ACG ACT GCT GAA TGG CGT
 A   V   N   K   S   V   E   E   T   T   A   E   W   R
AAA TTA TTA GCC GTC AAC CTT GAT GGT GTC TTC TTC GGT ACC CGA TTA
 K   L   L   A   V   N   L   D   G   V   F   F   G   T   R   L
GGG ATT CAA CGG ATG AAG AAC AAA GGC TTA GGG GCT TCC ATC ATC
 G   I   Q   R   M   K   N   K   G   L   G   A   S   I   I
AAC ATG TCT TCG ATC GAA GGC TTT GTG GGT GAT CCT AGC TTA GGG GCT
 N   M   S   S   I   E   G   F   V   G   D   P   S   L   G   A
TAC AAC GCA TCT AAA GGG GCC GTA CGG ATT ATG TCC AAG TCA GCT
 Y   N   A   S   K   G   A   V   R   I   M   S   K   S   A
GCC TTA GAT TGT GCC CTA AAG GAC TAC GAT GTT CGG GTA AAC ACT GTT
 A   L   D   C   A   L   K   D   Y   D   V   R   V   N   T   V
CAC CCT GGC TAC ATC AAG ACA CCA TTG GTT GAT GAC CTA CCA GGG
 H   P   G   Y   I   K   T   P   L   V   D   D   L   P   G
GCC GAA GAA GCG ATG TCA CAA CGG ACC AAG ACG CCA ATG GGC CAT
 A   E   E   A   M   S   Q   R   T   K   T   P   M   G   H
ATC GCT GAA CCT AAC GAT ATT GCC TAC ATC TGT GTT TAC TTG GCT TCT
 I   G   E   P   N   D   I   A   Y   I   C   V   Y   L   A   S
AAC GAA TCT AAA TTT GCA ACG GGT TCT GAA **TTC GTA GTT GAC GGT GGC**
 N   E   S   K   F   A   T   G   S   E   F   V   V   D   G   G
TAC ACT GCT CAA
 Y   T   A   Q

Fig. 4. DNA and protein sequence of the recombinant (R)-alcohol dehydrogenase from *Lactobacillus brevis* in *E. coli* (*upper line*: DNA sequence; *lower line*: corresponding amino acid in one letter code). The sequences of the primers are given in *bold-type*, sequences obtained by amino acid sequencing are *underlined*

studies on 3α,20β-hydroxysteroid dehydrogenase of *Streptomyces hydrogenans* revealed that the conserved tyrosine is located near the pyridine ring of the coenzyme and the conserved lysine is directly behind the tyrosine. Chemical modification and site-directed mutagenesis of short-chain ADHs demonstrate the involvement of this strictly conserved tyrosine [181–184]. From photoaffinity labelling studies on estradiol 17β-dehydrogenase it can be assumed that the C-terminal part of the protein is responsible for the substrate specificity of short-chain dehydrogenases [185].

The N-terminal sequences and the molecular mass of the subunits of the (R)-ADH from *L. kefir* and *L. brevis* clearly demonstrate that these new ADHs belong to the subgroup of short-chain dehydrogenases. Within this group these new dehydrogenases are the first examples of metal containing and metal requiring enzymes. Further studies concerning the structure-function relationship of these parts of the enzyme are necessary to elucidate the role of the cations.

3 Alcohol Dehydrogenases as Technical Catalysts

The biochemical characterization of several alcohol dehydrogenases and their exploitation potential demonstrate that these enzymes are most important tools for biochemists. Amino acid sequences of several ADHs are available so far, and alignment studies allow to establish ADH families and to consider their probable evolutionary relationships. For preparative applications, however, particular properties of an enzyme are essential prerequisites, such as enzyme stability and availability, its substrate specificity, or reaction selectivity. Enzymes with NAD as coenzyme are clearly preferred to NADP-dependent ones in practice, because NAD has a significantly higher stability [186–188], a lower price and, is in general, easier to regenerate.

3.1 Whole Cell vs. Enzymes as Means for the Reduction of Ketones

Alcohol dehydrogenases can be applied for the synthesis of chiral hydroxy compounds, either as whole cells or as a cellfree, more or less purified enzyme extract. As discussed in the introduction, whole cells are easy to use; coenzyme regeneration can simply be performed by the metabolic pathways of the cells, which can be fed by adding a degradable carbon source. Furthermore, the expensive isolation of the alcohol dehydrogenase can be avoided. However, there are several disadvantages: cells frequently contain enzymes that interfere with the desired reaction, so that a low yield and low enantiomeric excess of the product can be the result. Nevertheless, transformations with bakers' yeast have found their place in organic syntheses. This catalyst can be applied almost without any microbiological or biochemical experiences and is readily available. Remarkably, all these applications are limited to reduction reactions, whereas isolated enzymes in fact can be used in both directions, demanding separate coenzyme regeneration methods, only. Oxidation of alcohols requires particular strains of microorganisms, such as *Xanthobacter autotrophicus* to oxidize 2-chloroethanol [189], *Pseudomonas* sp., which oxidizes alkan-2-ols [190]; or methylotrophs [191–196].

In general, syntheses with isolated enzymes can be performed with higher selectivity and space-time yield than with whole cells, but they require in any case the coupling of coenzyme regenerating reactions.

3.2 Regeneration of Nicotinamide Coenzymes

Generally, the nicotinamide coenzymes are not covalently bound to the enzyme. They are employed in enzyme assays and preparative applications by adding catalytical but optimized amounts, and they need to be recycled. For an economic process, an efficient regeneration method is a basic requirement. The necessary recycle number depends essentially on the value of the chiral product, generally the method should recycle the coenzyme 100–100,000 times ([42]).

Numerous systems have been developed so far for the regeneration of the reduced coenzymes [197]. One of these is a chemical method with dithionite [198, 199], which is inexpensive, but unstable and which may reductively inactivate enzymes. Photochemical methods for coenzyme regeneration require photosensitizer such as tris(bipyridine)ruthenium(II) [200, 201] or *meso*-tetramethylpyridinium-porphyrinzinc(II) [202]. Electrochemical regeneration requires a suitable mediator, because direct cathodic reduction of NAD(P) generates undesirable byproducts of the coenzymes. Methylviologen has been widely used for mediator, but the reduced form requires a reductase for the reduction of NAD(P). Thus, this method is limited to the use of whole cells (electromicrobial reductions), that usually contain methylviologen-dependent NAD(P) reductases [154, 155, 203–205]. A non-enzymatic regeneration system for NADH and NADPH with a rhodium-bipyridinium complex as the redox catalyst and either the electrode or formate as the donor was developed recently by Steckhan et al. [206–208]. The formate-driven regeneration of NADH and NADPH was demonstrated for the stereoselective reduction of 4-phenyl-2-butanone with several alcohol dehydrogenases (HLADH, TBADH, (S)-ADH from *Rhodococcus erythropolis* and (R)-ADH from *Lactobacillus kefir*) [162]. Each system results in 4-phenyl-2-butanol with an enantiomeric excess of more than 96%. A water-soluble high-molecular weight derivative of this rhodium complex [209] facilitates continuous reduction reactions.

The most convenient and useful enzymatic methods for the regeneration of NAD(P)H are formate/formate dehydrogenase for NADH [210, 211], isopropanol/TBADH for NADPH [57], isopropanol/ADH (*Pseudomonas* sp.) for NADH [61, 212] and glucose/glucose dehydrogenase (*Bacillus* sp.) for NADH and NADPH [213].

For the regeneration of NADH, formate and formate dehydrogenase are most widely used. The enzyme from *Candida boidinii* is commercially available, formate is inexpensive, and the equilibrium favours a nearly irreversible reaction producing CO_2, and this product can easily be separated from the product solution. Disadvantages are the low specific activity of formate dehydrogenase (3–4 U/mg) and its price. This enzyme was already applied in large-scale processes using a continuously working enzyme-membrane-reactor. For this device, NAD was used as a molecular-weight enlarged derivative, which was obtained by chemical covalent binding of NAD to high-molecular weight polyethyleneglycole. The applicability of this technology was demonstrated by several applications in the field of enantioselective synthesis of hydroxy and amino acids (for a review see [2, 3]). Up to 600,000 moles of product may be produced per mole of NAD in such processes. Further limitation of the commercially available formate dehydrogenase arises from the fact that this enzyme exclusively reacts with NAD. Quite recently however, a formate dehydrogenase has been developed, which accepts both NAD and NADP [68, 69]. It could be achieved to broaden the coenzyme specificity by multipoint site-directed mutagenesis of the gene encoding for the NAD-dependent formate dehydrogenase in *Pseudomonas* sp. [214]. The activity of the mutant enzyme with NADP is in the range of 60% as compared to the NAD reduction. This enzymatic system may be applied in a continuous enzymatic reduction process to reduce acetophenone with the NADP-dependent ADH from *Lactobacillus* sp. [67].

The regeneration of the oxidized coenzymes NAD(P) is required for the synthesis of ketones from the racemic mixture of the corresponding hydroxy compounds. Besides the synthesis of enantiomerically pure hydroxyketones as the oxidation product of diols, this kind of reaction is important if ketones as the starting material are unstable or difficult to prepare. As a prerequisite to produce enantiomerically pure hydroxy compounds in this way the oxidation reaction must proceed completely. Methods for the regeneration of the oxidized nicotinamide coenzymes are less well developed than for the reduced coenzymes. Basically, regeneration of the oxidized coenzymes is hampered by the unfavourable thermodynamics and also by product inhibition [141]. Nevertheless, there are enzymatic methods for the regeneration based on α-ketoglutarate/L-glutamate dehydrogenase [141, 215–218], pyruvate or glyoxylate/lactate dehydrogenase [219] or acetaldehyde/ethanol dehydrogenase [220, 221]. A simple non-enzymatic method with flavin mononucleotide (FMN) was developed by Jones and Taylor [222–224]. FMN oxidizes NADH and FMN is afterwards regenerated by O_2.

In fact, the α-ketoglutarate/glutamate dehydrogenase is a generally applicable method for the regeneration of NAD and NADP in laboratory scale productions. Both components involved are inexpensive and stable. Quite recently, a method for the oxidation of the reduced nicotinamide coenzymes based on bacterial NAD(P)H oxidase has been described [225]. This enzyme oxidizes NADH as well as NADPH with low K_m values. The product of this reaction is peroxide, which tends to deactivate enzymes, but it can be destroyed simultaneously by addition of catalase. The irreversible peroxide/catalase reaction favours the ADH catalyzed oxidation reaction, and complete conversions of this reaction type are described.

3.3 Dehydrogenases in Organic Solvents

Substrates of alcohol dehydrogenases very often are almost insoluble in water. On the other hand, enzymes are generally isolated in aqueous buffers, and catalysis performed in aqueous solutions. Although enzymes as a rule are destroyed by the addition of organic especially water-miscible solvents, it has become evident in the last years, that enzymes may be active in the presence of organic solvents [226–230]. This was confirmed for enzymes that work originally in a hydrophobic environment, for instance in a biological membrane, as well as for very stable enzymes, such as proteases or enzymes from thermophilic organisms. The stability of enzymes in organic solvents depends on the hydrophobicity of the solvent, which can be expressed as the log P-value [231, 232]. Enzymes are more stable in nonpolar solvents with a low solubility for water than in polar solvents. In general, enzymes need a however small amount of water to retain a thin water layer on the surface, which is necessary to maintain the catalytically active conformation [233–237]. Coenzyme-dependent enzymes are rarely used in organic solvents. They are even less stable and less active in organic solvents than enzymes which require no cofactors; moreover, the coenzyme regeneration step certainly is an additional complication for such applications. Being a prerequisite for the application of alcohol dehydrogenases in organic solvents, their activity and stability were tested in the presence of nonpolar solvents (Table 15). These studies indicate that ADHs remain stable in the presence of organic solvents with a log P value of 3.5 or higher, solely YADH is unstable in all organic solvents tested so far. However, no results of preparative applications of ADHs in organic solvents are available until now.

Table 15. Influence of organic solvents on activity and stability of alcohol dehydrogenases. Values for the stability are given as the residual activity after 5-h contact with the solvents. The solvents are characterized and classified according to their log P-value [231]. Enzymes are active or stable in the presence of solvents at the mentioned log P value or higher. (Solvents in brackets are those that correspond to the mentioned log P-value.)

Alcohol dehydrogenase	Influence of Organic Solvents on	
	Stability	Activity
Lactobacillus kefir	log P > 2.5 (Toluol)	log P > 2.5 (Toluol)
Rhodococcus erythropolis	log P > 4 (Heptane)	log P > 3.5 (Hexane)
HLADH	log P > 1.2 (t-Butylmethylether)	log P > 1.2 (t-Butylmethylether)
TBADH	log P > 1.5 (Diisopropylether)	log P > 3 (Trichlorethan)
YADH	all solvents deactivate the enzyme completely (tested up to log P = 6.6 (Dodecan))	

Table 16. Chiral alcohols produced by continuous enzyme-catalyzed processes. The corresponding ketones are reduced with (S)-ADH from *Rhodococcus erythropolis*, NADH was regenerated by simultaneous coupling with formate dehydrogenase from *Candida boidinii* (FDH) and formate (data from [159])

Product alcohol	Ketone concentration [mM]	Space-time yield $[g \cdot L^{-1} \cdot d^{-1}]$	Enzyme consumption $[U \cdot kg \, (Product)^{-1}]$	
			(S)-ADH	FDH
(S)-1-Phenylpropan-2-ol	9.0	64	3540	10200
(S)-4-Phenylbutan-2-ol	12.0	104	3025	4860
(S)-6-Methylhept-5-en-2-ol	10.0	60	n.d.	n.d.

A more promising approach for the synthesis of hydrophobic substances with ADHs is published by Kruse et al. [159, 238]. They use a continuously operating reactor where the enzyme containing water phase is separated from the hydrophobic substrate-containing organic phase by a membrane. The hydrophobic product is extracted continuously via a hydrophobic membrane into an hexane phase, whereas the coenzyme is regenerated in a separate cycle, that consists of a hydrophilic buffer system. This method decouples advantageously the residence time of the cofactor from the residence time of the substrate. Several hydrophobic alcohols were prepared in this way with (S)-ADH from *Rhodococcus erythropolis* (Table 16).

3.4 Dehydrogenase-Catalyzed Preparation of Chiral Alcohols

Methods to produce chiral alcohols with ADHs are essentially described at a laboratory scale, namely those using HLADH, TBADH and the recently isolated enzymes from *Rhodococcus erythropolis*, and *Lactobacillus kefir* and *L. brevis*, respectively.

HLADH was applied for the reduction of a broad variety of substrates. The regeneration of NADH can be achieved by the ethanol coupled-substrate method [239]. Cyclic ketones are good substrates for this enzyme [137, 240, 241]. Heterocyclic compounds containing oxygen or sulfur are accepted, too [242–246], but nitrogen-containing substrates are inhibitors of HLADH, because they apparently complex with zinc in the active site of the enzyme [247].

Enantioselective reduction reactions with HLADH were carried out for instance with 2- and 3-keto esters [213, 248], *cis* and *trans* decaindiones [121] or cage shaped molecules [249].

As for oxidation reactions catalyzed by HLADH, the most frequently reported method is the coupling with FMN [222–224]. It has been used for instance for the oxidation of many meso-diols forming lactones [250, 251], or for the oxidation of primary alcohols to obtain chiral aldehydes [252]. Generally, these syntheses were carried out at 1–2 g scale within a reaction time of a few hours up to 2–3 weeks [247].

The NADP-dependent TBADH was used for the laboratory-scale preparation of several chiral aliphatic and cyclic hydroxy compounds by reduction of the corresponding ketones. For the regeneration of NADPH, this reduction reaction can be coupled with the TBADH catalyzed oxidation of isopropanol. For the reduction of some ketones it was observed that the reaction rate was increased in the presence of the regenerating substrate isopropanol, for instance in the presence of 0.2 v/v isopropanol, the reduction rate of butanone or pentanone was increased 3–4-fold [57]. In some cases, the enantiomeric excess of the reduction reaction is not very high, especially when small molecules are converted, but also for compounds such as acetophenone [138].

(R)-alcohols in high enantiomeric excess can be obtained with the aid of the NADP-dependent ADH from *Lactobacillus kefir*. Due to the broad substrate specificity of this enzyme, aromatic, cyclic, polycyclic as well as aliphatic ketones can be reduced. A simple method for the regeneration of NADPH is given by the simultaneously coupled oxidation of isopropanol by the same enzyme. Several chiral alcohols (Table 8) were synthesized at a 2.5 mmol scale within a reaction time of 12–36 h [160].

The preparation of some (S)-alcohols by ADH from *Rhodococcus erythropolis* has been described quite recently. This was the first report of a continuous production process for hydrophobic compounds. An important prerequisite of this method is a membrane which is resistant to organic solvents. It separates the hydrophilic phase, which contains the enzyme and the coenzyme, from the hydrophobic phase with the substrate and the product. Several products were prepared with this enzyme at a multigram scale (Table 16).

In summary, there are several alcohol dehydrogenases available for the synthesis of chiral alcohols. Most of the described enzymes belong to one of three groups which are precisely characterized. For preparative applications, only a few of these enzymes can be used, for their availability, stability, or substrate specificity may be insufficient. Only HLADH, TBADH, (R)-ADH from *Lactobacillus kefir* or *L. brevis* and the (S)-ADH from *Rhodococcus erythropolis* were found to be enzymes for general use. Until now, preparations of chiral alcohols have been carried out in laboratory-scale only; no product has been produced so far in a kilogram- or ton-scale by alcohol dehydrogenases. The low productivity of some of these enzymes surely means a limitation to the applicability. For instance, HLADH catalyzes the conversion of 1 mmol alcohol/day with about 20 mg of enzyme. Another drawback are the economics of this enantioselective step. For the conversion of water-soluble substrates such as keto acids, the product related enzyme consumptions are in the range of 500–2000 U/kg of product (Table 16). Hydrophobic substances, however, require significantly higher amounts of enzyme, which are in the range of 3000 U/kg for the ADH and 5000–10,000 U/kg for the formate dehydrogenase mediated regeneration step [159]. Enzyme costs of 500–1000 \$/1000 U for some ADHs and similar costs for formate dehydrogenase must hence be compared to alternate non-biological processes for the production of chiral alcohols.

Acknowledgement: This work was supported in part by the *Deutsche Forschungsgemeinschaft (DFG), Sonderforschungsbereich 380*: "Asymmetric syntheses with chemical and biological methods". I also wish to thank Bettina Riebel for her expert studies concerning the (*R*)-ADHs from *Lactobacillus* sp. and (*S*)-ADH from *Rhodococcus erythropolis* as well as for many stimulating discussions.

4 References

1. *Enzyme Nomenclature* (1979) New York: Academic Press
2. Hummel W, Kula M-R (1989) Eur J Biochem 184: 1
3. Kula M-R, Wandrey C (1987) Meth Enzymol 136: 9
4. Kragl U, Vasic-Racki D, Wandrey C (1993) Indian J Chem 32B: 103
5. Kim MJ, Whitesides GM (1988) J Am Chem Soc 110: 2959
6. Hogan JK, Parris W, Gold M, Jones JB (1992) Bioorg Chem 20: 204
7. Hummel W, Schütte H, Kula M-R (1983) Europ J Appl Microbiol Biotechnol 18: 75
8. Simon ES, Plante R, Whitesides GM (1989) Appl Biochem Biotechnol 22: 169
9. Bradshaw CW, Wong C-H, Hummel W, Kula M-R (1991) Bioorganic Chemistry 19: 29
10. Kim M-J, Kim JY (1991) J Chem Soc, Chem Commun : 326
11. Hummel W, Schütte H, Kula M-R (1985) Appl Microbiol Biotechnol 21: 7
12. Schütte H, Hummel W, Kula M-R (1984) Appl Microbiol Biotechnol 19: 167
13. Hummel W, Schütte H, Kula M-R (1988) Appl Microbiol Biotechnol 28: 433
14. Vasic-Racki D, Jonas M, Wandrey C, Hummel W, Kula M-R (1989) Appl Microbiol Biotechnol 31: 215
15. Bommarius AS, Drauz K, Hummel W, Kula M-R, Wandrey C, Biocatalysis 10: 37
16. Bommarius A, Schwarm M, Stingl K, Kottenhan M, Huthmacher K, Drauz K (1995) Tetrahedron: Asymmetry 6: 2851
17. Wandrey C, Bossow B (1986) Biotechnol Bioind 3: 8
18. Hanson RL, Singh J, Kissik TP, Patel RN, Szarka LJ, Mueller RH (1990) Bioorg Chem 18: 116
19. Hummel W, Schütte H, Schmidt E, Wandrey C, Kula M-R (1987) Appl Microbiol Biotechnol 26: 409
20. Seebach D (1990) Angew Chem 102: 1363
21. Nikolova P, Ward OP (1992) Biotechnol Bioeng 39: 870
22. Ward OP, Young CS (1990) Enzyme Microb Technol 12: 482
23. Servi S (1990) Synthesis : 1
24. Csuk R, Glänzer B (1991) Chem Rev 91: 49
25. Sih CJ, Chen C-S (1984) Angew Chem Int Ed Engl 23: 570
26. Buisson D, Azerad R, Sanner C, Larcheveque M (1992) Biocatalysis 5: 249
27. Seebach D, Roggo S, Maetzke T, Braunschweiger H, Cercus J, Krieger M (1987) Helv Chi Acta 70: 1605
28. Sheih W-R, Gopalan AS, Sih CJ (1985) J Am Chem Soc 107: 2993
29. Bernardi R, Ghiringhelli D (1987) J Org Chem 52: 5021
30. Bernardi R, Cardillo R, Ghiringhelli D, de Pavo V (1987) J Chem Soc Perkin Trans I: 1607
31. Ushio K, Inouye K, Nakamura K, Oka S, Ohno A (1986) Tetrahedron Lett 27: 2657
32. Nakamura K, Inoue K, Ushio K, Oka S, Ohno A (1987) Chem Lett : 679
33. Nakamura K, Kawai Y, Oka S, Ohno A (1989) Bull Chem Soc Jpn 62: 875
34. Ushio K, Ebara K, Yamashita T (1991) Enzyme Microb Technol 13: 834
35. Chibata I, Tosa T, Sato T (1974) Appl Microbiol 27: 878
36. Nakamura K, Higaki M, Ushio K, Oka S, Ohno A (1985) Tetrahedron Lett : 4213
37. Prelog V (1964) Pure Appl Chem 9: 119
38. Jones JB, Jakovac IJ (1982) Can J Chem 60: 19
39. Pereira RS (1995) Appl Biochem Biotehcnol 55: 123
40. Kroner K-H, Schütte H, Stach W, Kula M-R (1982) J Chem Tech Biotechnol 32: 130
41. Cordes A, Kula M-R (1986) J Chromatography 376: 375
42. Wong C-H, Whitesides GM. 1994. *Enzymes in synthetic organic chemistry*. New York: Pergamon, Elsevier Science Inc

43. Faber K. 1995. *Biotransformations in organic chemistry*. Berlin, Heidelberg, New York: Springer-Verlag
44. Davies HG, Green RH, Kelly DR, Roberts SM. 1989. *Biotransformations in preparative organic chemistry*. London: Academic Press
45. Morrison JD. 1985. *Chiral Catalysis*. London: Academic Press
46. Sheldon JD. 1993. *Chirotechnology*. New York: Marcel Dekker
47. Millership JS, Fitzpatrick A (1993) Chirality 5: 573
48. Hummel W. (1997) In *Frontiers in Biosensorics*, ed. F. Scheller, J. Fedrowitz, F. Schubert. Basel: Birkhäuser, p. 47
49. Goodhue CT. 1982. In *Microbial transformations of bioactive compounds*, Vol. I, ed. J.P. Rosazza, pp. 9. Boca Raton, Florida: CRC Press
50. Elander RP. 1987. In *Basic Biotechnology*, ed. J. Bu'Lock, B. Kristiansen, pp. 217. London: Academic Press
51. Cheetham PSJ (1987) Enz Microb Technol 9: 194
52. Hummel W, Weiss N, Kula M-R (1984) Arch Microbiol 137: 47
53. Hummel W, Schmidt E, Wandrey C, Kula M-R (1986) Appl Microbiol Biotechnol 25: 175
54. Hummel W, Schmidt E, Schütte H, Kula M-R. 1987. In *Proceedings Biochemical Engineering*, ed. H.-J. Chmiel, pp. 392. Stuttgart: Fischer Verlag
55. Hummel W, Wendel U, Sting S. 1992. In *Biosensors: Fundamentals, Technologies and Applications.*, ed. F. Scheller, R.D. Schmid, pp. 381. Weinheim: VCH Publisher
56. Schneider KH, Jakel G, Hoffmann R, Giffhorn F (1995) Microbiology 141: 1865
57. Lamed R, Keinan E, J.G. Z (1981) Enzyme Microb Technol 3: 144
58. Guagliardi A, Martino M, Iaccarino I, Derosa M, Rossi M, Bartolucci S (1996) Int J Biochem Cell Biol 28: 239
59. Rella R, Raia CA, Pensa M, Pisani FM, Gambacorta A, de Rosa M, Rossi M (1987) Eur J Biochem 167: 475
60. Ammendola S, Raia CA, Caruso C, Camardella L, Dauria S, Derosa M, Rossi M (1992) Biochemistry 31: 12514
61. Bradshaw CW, Fu H, Shen G-J, Wong C-H (1992) J Org Chem 57: 1526
62. Bradshaw C, Shen G-J, Wong C-H (1991) Bioorg Chem 19: 398
63. Suzuki Y, Oishi K, Nakano H, Nagayama T (1987) Appl Microbiol Biotechnol 26: 546
64. Wilks HM, Hart KW, Freeney R, Dunn CR, Muirhead H, Chia WN, Barstow DA, Atkinson T, Clarkee AR, Holbrook JJ (1988) Science 242: 1541
65. Luyten MA, Bur D, Wynn H, Parris W, Gold M, Friesen JD, Jones JB (1989) J Am Chem Soc 111: 6800
66. Corbier C, Clermont S, Billard CP, Skarzynski T, Branlant C, Wonacott A, Branlant G (1990) Biochemistry 29: 7101
67. Seelbach K, Riebel B, Hummel W, Kula M-R, Tishkov VI, Egorov AM, Wandrey C, Kragl U (1996) Tetrahedron Lett 37: 1377
68. Tishkov VI, Galkin AG, Kulakova LB, Egorov AM. 1993. In *Enzyme Engineering XII*. Deauville, France:
69. Tishkov VI, Galkin AG, Marchenko GN, Tsygankov YD, Egorov AM (1993) Biotechnol Appl Biochem 18: 201
70. Cronin CN, Malcolm BA, Kirsch JF (1987) J Am Chem Soc 109: 2222
71. Murali C, Creaser EH (1986) Protein Engineering 1: 55
72. Zhong Z, Liu JL-C, Dinterman LM, Finkelman MAJ, Mueller WT, Rollence ML, Whitlow M, Wong C-H (1991) J Am Chem Soc 113: 683
73. Bonneau PR, Graycar TP, Estell DA, Jones JB (1991) J Am Chem Soc 113: 1026
74. Pantoliano MW, Whitlow M, Wood JF, Dodd SW, Hardman KD, Rollence ML, Bryan PN (1989) Biochemistry 28: 7205
75. Russell AJ, Fersht AR (1987) Nature : 328
76. Russell AJ, Fersht AR (1987) Nature : 496
77. Fersht A, Winter G (1992) TIBS : 92
78. Wells JA, Estell DA (1988) Trends Biochem Sci 13: 291
79. Jörnvall H, Persson B, Jeffery J (1987) Eur J Biochem 167: 195
80. Reid MF, Fewson CA (1994) Crit Rev Microbiol 20: 13
81. Persson B, Krook M, Jörnvall H (1991) Eur J Biochem 200: 537
82. Persson B, Zigler JS, Jörnvall H (1994) Eur J Biochem 226: 15
83. Schauder S, Schneider K-H, Giffhorn F (1995) Microbiology 141: 1857

84. MacKintosh RW, Fewson CA (1988) Biochem J 250: 743
85. Mallinder PR, Pritchard A, Moir A (1992) Gene 110: 9
86. Hummel W (1990) Appl Microbiol Biotechnol 34: 15
87. Lamed RJ, Zeikus JG (1981) Biochem J 195: 183
88. Lamed RJ, Zeikus JG (1980) J Bacteriol 141: 1251
89. Conway T, Ingram LO (1989) J Bacteriol 171: 3754
90. Riebel B, Hummel W. 1996 pers. commun.
91. Saliola M, Shuster JR, Falcone C (1990) Yeast 6: 193
92. Bränden C-I, Jörnvall H, Eklund H, Furugren B. 1975. In *The Enzymes*, Vol. 11, ed. PD Boyer, pp. 103. New York: Academic Press
93. Jörnvall H, von Bahr-Lindström H, Jeffery J (1984) Eur J Biochem 140: 17
94. Riebel B. 1996. Thesis, University of Düsseldorf, Düsseldorf
95. Jörnvall H, Persson B, Jeffery J (1987) Eur J Biochem 167: 195
96. Borrás T, Persson B, Jörnvall H (1987) Eur J Biochem 28: 6133
97. Aronson BD, Somerville RL, Epperly BR, Dekker EE (1989) J Biol Chem 264: 5226
98. Karlsson C, Maret W, Auld DS, Höög J-O, Jörnvall H (1989) Eur J Biochem 186: 543
99. Bright JR, Byrom D, Danson MJ, Hough DW, Towner P (1993) Eur J Biochem 211: 549
100. Amy CM, Witkowski A, Naggert J, Williams B, Randhawa Z, Smith S (1989) Proc Natl Acad Sci USA 86: 3114
101. Donadio S, Staver MJ, McAlpine JB, Swanson SJ, Katz L (1991) Science 252: 675
102. Eklund H, Nordström B, Zeppezauer E, Söderlund G, Ohlsson I, Boiwe T, Brändén C-I (1974) FEBS Lett 44: 200
103. Eklund H, Nordström B, Zeppezauer E, Söderlund G, Ohlsson I, Boiwe T, Söderberg B-O, Tapla O, Bränden C-I (1976) J Mol Biol 102: 27
104. Eklund H, Samama J-P, Wallen L, Brändén C-I (1981) J Mol Biol 146: 561
105. Cedergen-Zeppezauer ES, Andersson I, Ottonello S (1985) Biochemistry 24: 4000
106. Jörnvall H, Eklund H, Branden CI (1978) J Biol Chem 253: 8414
107. Ganzhorn AJ, Green DW, Hershey AD, Gould RM, Plapp BV (1987) J Biol Chem 262: 3754
108. Jörnvall H, von Bahr-Lindström H, Jany K-D, Ulmer W, Fröschle M (1984) FEBS Lett 165: 190
109. Jörnvall H, Persson M, Jeffery J (1981) Proc Natl Acad Sci USA 78: 4226
110. Thatcher DR, Sawyer L (1980) Biochem J 187: 884
111. Villarroya A, Juan E, Egestad B, Jörnvall H (1989) Eur J Biochem 180: 191
112. Dothie JM, Giglio JR, Moore CB, Taylor SS, Hartley BS (1985) Biochem J 230: 569
113. Taylor SS, Rigby PJW, Hartley BS (1974) Biochem J 141: 693
114. Pauly HE, Pfleiderer G (1975) Hoppe-Seylers Z Physiol Chem 356: 1613
115. Jany KD, Ulmer W, Froschle M, Pfleiderer G (1984) FEBS Lett 165: 6
116. Conway T, Sewell GW, Osman YA, Ingram IO (1987) J Bacteriol 169: 2591
117. Williamson VM, Paquin CE (1987) Mol Gen Genet 209: 374
118. Chou PY, Fasman GD (1974) Biochemistry 13: 222
119. You KS (1984) Crit Rev Biotechnol 17: 313
120. Arnold LJ, You K, Allison WS, Kaplan NO (1976) Biochemistry 15: 4844
121. Dodds DR, Jones JB (1988) J Am Chem Soc 110: 577
122. Nakazaki M, Chikamatsu H, Naemura K, Hirose Y, Shimizu T, Asao M (1978) J Chem Soc Chem Commun : 667
123. Nakazaki M, Chikamatsu H, Naemura K, Asao M (1980) J Org Chem 45: 4432
124. Nakazaki M, Chikamatsu H, Naemura K, Suzuki T, Iwasaki M, Sasaki Y, Fujii T (1981) J Org Chem 46: 2726
125. Nakazaki M, Chikamatsu H, Fujii T, Susaki Y, Ao S (1983) J Org Chem 48: 4337
126. Hansch C, Bjorkroth (1986) J Org Chem 51: 5461
127. Dutler H, Branden C-I (1981) Bioorg Chem 10: 1
128. Horjales E, Branden C-I (1985) J Biol Chem 260: 15445
129. Lemière GL, van Osselaer TA, Lepoivre JA, Alderweireldt FC (1982) J Chem Soc Perkin Trans II : 1123
130. Kuassman J, Petterson G (1980) Eur J Biochem 103: 557
131. Klinman JP (1981) Crit Rev Biochem : 39
132. MacLeod R, Prosser H, Fikentscher L, Lanyi J, Mosher HS (1964) Biochemistry 3: 838
133. Nakamura K, Miyai T, Kawai J, Nakajima N, Ohno A (1990) Tetrahedron Lett : 1159
134. Wong C-H, Whitesides GM (1983) J Am Chem Soc 105: 5012
135. Mansson MO, Larsson PO, Mosbach K (1982) Meth Enzymol 89: 457

136. Wang SS, King C-K (1979) Adv Biochem Eng 12: 119
137. Dodds DR, Jones JB (1982) J Chem Soc, Chem Commun : 1080
138. Keinan E, Hafeli EK, Seth KK, Lamed R (1986) J Am Chem Soc 108: 162
139. Keinan E, Seth KK, Lamed R, Ghirlando R, Singh SP (1990) Biocatalysis 3: 57
140. Keinan E, Sinha SC, Sinha-Bagchi A (1991) J Chem Soc, Perkin Trans 1: 3333
141. Lee LG, Whitesides GM (1986) J Org Chem 51: 25
142. Carrea G, Bovara R, Longhi R, Riva S (1985) Enzyme Microb Technol 7: 597
143. Carrea G (1984) Trends Biotechnol 2: 102
144. Carrea G, Bovara R, Longhi R, Barani R (1984) Enzyme Microbiol Technol 6: 307
145. Hummel W, Boermann F, Kula M-R (1989) Biocatalysis 2: 293
146. Hata H, Shimizu S, Hattori S, Yamada H (1990) J Org Chem 55: 4377
147. Peters J, Zelinski T, Minuth T, Kula M-R (1993) Tetrahedron Asymmetry 4: 1173
148. Peters J, Minuth T, Kula M-R (1993) Enzyme Microb Technol 15: 950
149. Shimizu S, Hattori S, Hata H, Yamada H (1988) Eur J Biochem 174: 37
150. Hata H, Shimizu S, Hattori H, Yamada H (1989) FEMS Microbiol Lett 58: 87
151. Hata H, Shimizu S, Hattori S, Yamada H (1989) Biochim Biophys Acta 990: 175
152. Shimizu S, Hata H, Yamada H (1984) Agric Biol Chem 48: 2285
153. Tischer W, Bader J, Simon H (1979) Eur J Biochem 97: 103
154. Simon H, Günther H, Bader J, Tischer W (1981) Angew Chem Int Ed Engl 20: 861
155. Simon H, Bader J, Günther H, Neumann S, Thanos J (1985) Angew Chem Int Ed Engl 24: 539
156. Keinan E, Seth KK, Lamed R (1987) Ann NY Acad Sci 501: 130
157. Hummel W (1990) Biotechnol Lett 12: 403
158. Hummel W, Goffwald, C. (1991) Ger. Pat. Appl. P 42.09.022.9
159. Kruse W, Hummel W, Kragl U (1996) Recl Trav Chim Pays-Bas 115: 239
160. Bradshaw CW, Hummel W, Wong C-H (1992) J Org Chem 57: 1532
161. Hughes MN, Poole RK. 1989. *Metals and micro-organisms.* London: Chapman and Hall
162. Westerhausen D, Herrmann S, Hummel W, Steckhan E (1992) Angew Chem Int Ed Engl 31: 1592
163. 1974. *Bergey's Manual of Determinative Bacteriology.* Baltimore: Williams & Wilkins
164. Orla S. 1919. *The lactic acid bacteria.* Copenhagen: Fred Host and Son
165. Orla S. 1943. *The lactic acid bacteria.* Copenhagen: Munksgaard
166. Rogosa M (1970) Int J System Bact 20: 519
167. Wilkinson BJ, Jones D (1977) J Gen Microbiol 98: 399
168. Williams RAD (1971) J Dent Res 50: 1104
169. Buyze G, van den Hamer CJA, de Haan PG (1957) J Microbiol Serol 23: 345
170. Delong A, Calderonurrea A, Dellaporta SL (1993) Cell 74: 757
171. Yoshimoto T, Hagashi H, Kanatani A, Lin XS, Nagai H, Oyama H, Kurazono K, Tsuru D (1991) J Bacteriol 173: 2173
172. Krook M, Marekov L, Jörnvall H (1990) Biochemistry 29: 738
173. Yamamoto-Otake H, Koyama Y, Horiuchi T, Nakano E (1991) Appl Environ Microbiol 57: 1418
174. Neidle E, Hartnett C, Ornston N, Bairoch A, Rekik M, Harayama S (1992) Eur J Biochem 204: 113
175. Hofer B, Eltis LD, Dowling DN, Timmis KN (1993) Gene 130: 47
176. Verwoert IIGS, Verbree EC, van der Linden KH, Nijkamp HJJ, Stuitje AR (1992) J Bacteriol 174: 2851
177. Yin SJ, Vagelopoulos N, Lundquist G, Jörnvall H (1991) Eur J Biochem 197: 359
178. Marekov L, Krook M, Jörnvall H (1990) FEBS Lett 266: 51
179. Tannin GM, Agarwal AK, Monder C, New MI, White PC (1991) J Biol Chem 266: 16653
180. Agarwal AK, Monder C, Eckstein B, White PC (1989) J Biol Chem 264: 18939
181. Albalat R, Gonzàlez-Duarte R, Atrian S (1992) FEBS Lett 308: 235
182. Ensor CM, Tai H-H (1991) Biochem Biophys Res Commun 176: 840
183. Chen Z, Jiang JC, Lin ZG, Lee WR, Baker ME, Chang SH (1993) Biochemistry 32: 3342
184. Krook M, Prozorovski V, Atrian S, Gonzàlez-Duarte R, Jörnvall H (1992) Eur J Biochem 209: 233
185. Murdock GL, Chin CC, Warren JC (1986) Biochemistry 25: 641
186. Wong C-H, Whitesides GM (1981) J Am Chem Soc 103: 4890
187. Johnson SL, Tuazon PT (1977) Biochemistry 16: 1175
188. Oppenheimer NJ, Kaplan NO (1974) Biochemistry 13: 4675
189. Janssen DB, Keuning S, Witholt B (1987) J Gen Micorbiol 133: 85

190. Rigby DJ, Dodgson KS, White GF (1986) J Gen Microbiol 132: 35
191. Steinbüchel A, Schlegel HG (1984) Eur J Biochem 141: 555
192. Bellion E, Wu GT-S (1978) J Bacteriol 135: 251
193. Yamanaka K, Minoshima R (1984) Agric Biol Chem 48: 171
194. Hou CT, Patel R, Laskin AI, Barnabe N, Marczak I (1979) Appl Environ Microbiol 38: 135
195. Hou CT, Patel R, Barnabe N, Marczak I (1981) Eur J Biochem 199: 359
196. Hiu SF, Zhu C-X, Yan R-T, Chen J-S (1987) Appl Environ Microbiol 53: 697
197. Chenault HK, Whitesides GM (1987) Appl Biochem Biotechnol 14: 147
198. Jones JB, Sneddon DW, Higgins W, Lewis AJ (1972) J Chem Soc Chem Commun: 856
199. Vandecasteele J-P, Lemal J (1980) Bull Soc Chim France 101
200. Wienkamp R, Steckhan E (1983) Angew Chem Int Ed Engl 22: 497
201. Mandler D, Willner I (1984) J Am Chem Soc 106: 5352
202. Mandler D, Willner I (1986) J Chem Soc, Perk Trans 2: 805
203. Bader J, Günther H, Nagata S, Schütz H-J, Link M-L, Simon H (1984) J Biotechnol 1: 95
204. Günther H, Frank C, Schütz H-J, Bader J, Simon H (1983) Angew Chem Int Ed Engl 22: 322
205. Day RJ, Kinsey SJ, Seo ET, Weliky N, Silverman HP (1972) Trans NY Acad Sci 34: 588
206. Steckhan E, Hermann S, Ruppert R, Dietz E, Frede M, Spika E (1991) Organometallics 10: 1568
207. Ruppert R, Herrmann S, Steckhan E (1987) Tetrahedron Lett 28: 6583
208. Ruppert R, Herrmann S, Steckhan E (1988) J Chem Soc Chem Commun: 1150
209. Steckhan E, Herrmann S, Ruppert R, Thömmes J, Wandrey C (1990) Angew Chem Int Ed Engl 29: 388
210. Wichmann R, Wandrey C, Bückmann AF, Kula M-R (1981) Biotechnol Bioeng 23: 2789
211. Shaked Z, Whitesides GM (1980) J Am Chem Soc 102: 7104
212. Shen G-J, Wang Y-F, Wong C-H (1990) J Chem Soc, Chem Commun 9: 677
213. Wong C-H, Drueckhammer DG, Sweers HM (1985) J Am Chem Soc 107: 4028
214. Egorov AM, Avilova TV, Dikov MM, Popov VO, Radionov YV, Berezin IV (1979) Eur J Biochem 99: 569
215. Wong C-H, Matos JR (1985) J Org Chem 50: 1992
216. Wong C-H, McCurry SD, Whitesides GM (1980) J Am Chem Soc 102: 7938
217. Lee LG, Whitesides GM (1985) J Am Chem Soc 107: 6999
218. Carrea G, Bovara R, Cremonesi P, Lodi R (1984) Biotechnol Bioeng 26: 560
219. Wong C-H, Whitesides GM (1982) J Org Chem 47: 2816
220. Lemière GL, Lepoivre JA, Alderweireldt FC (1985) Tetrahedr Lett 26: 4527
221. Chambers RP, Walle EM, Baricos WH, Cohen W (1978) Enz Eng 3: 363
222. Jones JB, Taylor KE (1973) Chem Comm : 205
223. Jones JB, Taylor KE (1976) Can J Chem 54: 2974
224. Jones JB, Taylor KE (1976) Can J Chem 54: 2969
225. Hummel W, Riebel B (1996) Enz Eng 13: 713
226. Cambou B, Klibanov AM (1984) J Am Chem Soc 106: 2687
227. Yokozeki K, Yamanaka S, Takinami K, Hirose Y, Tanaka A, Sonomoto K, Fukui S. (1982) Eur J Appl Microbiol Biotechnol 14: 1
228. Martinek K, Semenov AN, Berezin IV (1981) Biocheim Biophys Acta 658: 76
229. Klibanov AM, Samokhin GP, Martinek K, Berezin IV (1977) Biotechnol Bioeng 19: 1351
230. Dastol FR, Musto NA, Price S (1966) Arch Biochem Biophys 115: 44
231. Laane C, Boeren S, Hilhorst RC, Veeger C. 1987. In Biocatalysis in Organic Media, ed. C. Laane, J. Tramper, M.D. Lilly, pp. 65. Amsterdam:
232. Laane C, Boeren S, Vos K, Voeger C (1987) Biotechnol Bioeng 30: 81
233. Zaks A, Klibanov AM (1988) J Biol Chem 263: 8017
234. Zaks A, Klibanov AM (1988) J Biol Chem 263: 3194
235. Schinkel JE, Downer NW, Rupley JA (1985) Biochemistry 24: 352
236. Luisi PL, Laane C (1986) TIBTECH : 153
237. Careri G, Gratton E, Yang P-H, Rupley JA (1980) Nature 284: 572
238. Kragl U, Kruse W, Hummel W, Wandrey C (1996) Biotechnol Bioeng (in press)
239. Zagalak B, Frey PA, Karabatsos GL, Abeles RH (1966) J Biol Chem 241: 3028
240. Krawczyk AR, Jones JB (1989) J Org Chem 54: 1795
241. Nakazaki M, Chikamatsu H, Taniguchi M (1982) Chem Lett : 1761
242. Jones JB, Takemura T (1984) Can J Chem 62: 77
243. Davies J, Jones JB (1979) J Am Chem Soc 101: 5405

244. Haslegrave JA, Jones JB (1982) J Am Chem Soc 104: 4666
245. Sadozai SK, Lepoivre JA, Dommisse RA, Alderweireldt FC (1980) Bull Soc Chim Belg 89: 637
246. Jones JB, Schwartz HM (1981) Can J Chem 59: 1574
247. Jones JB. 1985. In *Enzymes in organic synthesis*, pp. 3. London: Pitman
248. Drueckhammer DG, Sadozai SK, Wong C-H, Roberts SM (1987) Enzyme Microb Technol 9: 564
249. Naemura K, Katoh T, Chikamatsu H, Nakasaki M (1984) Chem Lett: 1371
250. Bridges AJ, Raman PS, Ng GSY, Jones JB (1984) J Am Chem Soc 106: 1461
251. Jakovac JJ, Goodbrand HB, Lok KP, Jones JB (1982) J Am Chem Soc 104: 4559
252. Yamazaki Y, Hosono K (1988) Tetrahedron Lett 29: 5769

Fluidized Bed Adsorption as a Primary Recovery Step in Protein Purification

Jörg Thömmes
Institut für Enzymtechnologie, Heinrich-Heine Universität Düsseldorf,
D-52404 Jülich, Germany, Email: thoemmes@ibt022.ibt.kfa-juelich.de

Dedicated to Professor Dr. Maria-Regina Kula on the occasion of her 60th birthday

Fluidized bed adsorption has been introduced as an integrative technology combining clarification, concentration, and initial purification in a single step. In the paper presented here, the use of fluidized adsorbents in the primary recovery of proteins starting from unclarified broths is reviewed. First the principle of fluidizing adsorbent particles is discussed, subsequently possible experimental procedures for whole broth adsorption are demonstrated. The system parameters governing the performance of the sorption process in a fluidized bed are discussed in the second part of the paper and considerations on how operating parameters and process design influence the limiting steps are provided. Finally, examples for the successful operation of whole broth adsorption employing fluidized adsorbents are shown and conditions are defined under which this technology may be an alternative to traditional protein purification methods.

List of Symbols and Abbreviations

α	proportionality constant in Eq. (19)
ε	interstitial volume
η	dynamic viscosity [kg/(m·s)]
ρ_l	liquid density [kg/m^3]
ρ_p	solids density [kg/m^3]
μ_1	first central moment of the residence time distribution
μ_2	second central moment of the residence time distribution
Bo	Bodenstein number $U \cdot L/D_{axl}$
Ga	Gallileo number $\rho_g g(\rho_p - \rho_l)d_p^3/\eta^2$
Pe_p	Particle Peclet number $U_e \cdot d_p/D_{axl}$
Re_p	Particle Reynolds number $U \cdot d_p \cdot \rho_l/\eta$
Re_{mf}	Reynolds number at minimum fluidization velocity
Re_t	Reynolds number at terminal setting velocity
Sc	Schmidt number $\eta/(D \cdot \rho_l)$
BSA	bovine serum albumin
CIP	cleaning in place
D	diffusion coefficient in free solution [m^2/s]
D_{axl}	liquid phase dispersion coefficient [m^2/s]
D_{axp}	solid phase dispersion coefficient [m^2/s]
D_e	effective (pore) diffusion coefficient [m^2/s]
d_c	column diameter [m]
d_p	particle diameter [m]
G6PDH	glucose-6-phosphate dehydrogenase
HSA	human serum albumin
IP	isoelectric point
K	equilibrium constant
k_f	film transport coefficient [m/s]
L	fluidized bed height [m]
m	proportionality constant in Eq. (19)
M_A	protein molecular weight [kDa]
MAb	monoclonal antibody

MDH	malate dehydrogenase
MSFB	magnetically stabilized fluidized bed
Mv	density ratio $(\rho_p - \rho_l)/\rho_l$
n	Richardson-Zaki coefficient
N_L	fluid side mass transport number
N_P	particle side mass transport number
Q_{dyn}	dynamic capacity $[\mathrm{mg/ml_{adsorbent}}]$
Q_{max}	equilibrium capacity $[\mathrm{mg/ml_{adsorbent}}]$
r_p	particle radius $[\mathrm{m}]$
SSRFB	single stage recirculating fluidized bed
T	absolute temperature $[\mathrm{K}]$
U	linear velocity $[\mathrm{m/s}]$
U_e	effective linear velocity (U/ε) $[\mathrm{m/s}]$
U_{mf}	minimum fluidization velocity $[\mathrm{m/s}]$
U_t	terminal settling velocity $[\mathrm{m/s}]$

1 Introduction

1.1 Primary Protein Recovery – Defining the Problem

Technologies suitable for the purification of proteins have to meet boundary conditions, which are different from conventional methods of downstream processing usually employed in chemical technology. This is on one hand due to the sensitivity of the protein macromolecule to temperature and interfacial contact as well as to any agent affecting its three-dimensional structure. The second important difference to a typical purification of chemically produced compounds is the medium, from which the protein has to be isolated. Typically, protein sources, e.g. fermentation broths or cell, plant, and tissue homogenates, contain particulate material, which has to be removed prior to further fractionation of the feedstock. The liquid itself contains a large variety of dissolved compounds in comparatively low concentrations, the desired protein often being only a minor component in this mixture. A concentration of the initial solution therefore is indicated in advance to further purification steps which may be large in number and of a complex nature. The complexity of the purification task is mirrored by the fact, that mostly chromatographic techniques are the only methods guaranteeing the desired selectivity of the separation. Concentration, e.g. by ultrafiltration or precipitation, as well as conventional protein chromatography is compromised by the presence of particulate material in the feed solutions. Therefore the first task within a protein purification procedure is usually solid-liquid separation which is supposed to deliver a clarified solution. The supernatant should then be concentrated followed by the sequence of operations mentioned above. The delivery of a clarified concentrate can be summarized under the category initial recovery, the further processing steps cover the categories of low and high resolution purification, as defined by Wheelwright [1].

From a point of view of industrial protein production the number of sequential operations necessary to achieve the desired purity of a protein contributes significantly to the overall costs of the downstream process. This is on one hand due to the capital investment and amount of consumables needed for each step as well as to the individual time requirements of each operation, as labour costs are a very important factor in the calculation of process economics. Secondly the overall yield of the purification is reduced with each additional process step, originating from its inherent loss of product. Furthermore, fast primary recovery should separate the protein of interest from process conditions detrimental to its structural stability, e.g. proteases, glycosidases, or oxidizing conditions. As the performance of the purification process, expressed by its overall yield, operation time, and capital cost may contribute to up to 80% of the total production costs [2], it is evident, that a reduction of the number of sequential steps in a purification protocol may be the key to the economic success of a potential protein product [3].

Fast and efficient primary recovery steps in advance of low and high resolution methods for further purification of proteins form the basis for a successful downstream process. Particularly on a large scale, these operations may pose a significant problem which is not evident to the biotechnologist performing protein purification on a bench scale. The standard laboratory protocols for clarification of culture broths or homogenates are easy to follow and may be performed comparatively quickly. Small volumes of particle-containing suspensions are simply spun down in high speed centrifuges and residual particles are removed by dead-end microfiltration. The simplicity of the operations give the impression that clarification of the initial broth will be no problem during scale up. Datar and Rosen [4] provided an excellent overview on the actual situation in large-scale cell and cell-debris removal. They showed, that centrifugal separators may be used very efficiently on a very large scale for the clarification of broths containing comparatively large particles, e.g. yeast cells, but that the performance is rapidly reduced with decreasing particle size (bacterial cells, debris). Additionally shear sensitive organisms such as mammalian cells may be damaged in the centrifugal field, findings which are supported by the data of Kempken et al. [5]. Crossflow filtration as an alternative method is well-established up to the production scale for eukaryotic [6] and prokaryotic [7] systems as well as for debris [4], but suffers from comparatively low steady state liquid flux rates due to fouling of the membranes by small particles, lipids, nucleic acids, and the like. High product concentrations in the culture broth, which today are achieved with modern methods of recombinant DNA technology, may also lead to irreversible adsorption of the protein of interest to the membrane, resulting in reduced product yield during clarification. In conclusion, Datar and Rosen underlined the necessity of developing novel solutions for clarification of biological suspensions containing very small particles such as cell debris.

Summarizing the points discussed above, primary recovery of proteins from biological suspensions by centrifugation and filtration is a well-established technology, which is successfully operated on a very large scale. There are, however, specific problems in removing small particles or sensitive organisms, so there is enough room for alternative technologies to prove their potential. Additionally there is a need to streamline downstream processes due to their large contribution to the overall costs of a protein production, so an integration of several recovery or purification steps into novel methods is indicated.

1.2 Integrative Protein Recovery

As a consequence of the discussion on the efficiency of conventional primary recovery of proteins, integrative operations, which on one hand simplify solid-liquid separation and on the other hand combine originally independent steps to new unit operations, seem to meet the criteria for possible improvements.

Integrative protein recovery operations are supposed to tolerate particle-containing biological suspensions as initial feedstock and to deliver a clarified product concentrate which can be transferred to further purification steps. Ideally, a first fractionation of the protein matrix contained in the feed is performed as well, thus combining clarification, concentration, and capture in a single process step.

There are three different approaches described in the literature which meet the criteria formulated above: liquid-liquid extraction based on aqueous two-phase systems allows processing of biomass-containing feedstocks at very high biomass loads. This procedure has been proven to work very efficiently for homogenates as well as for whole cell broths by replacing the mechanical solid-liquid separation in a centrifugal separator by a thermodynamically controlled partitioning of particulate material and dissolved protein product in two different aqueous phases [8, 9]. By choosing systems with suitable differences in the physicochemical properties of the phases an initial fractionation of the protein matrix may be achieved with very good selectivity. Introducing protein binding ligands to crossflow microfiltration membranes allows the combination of filtration and protein adsorption. Kroner et al. have investigated the feasibility of this concept for the purification of enzymes from unclarified homogenates [10]. The third class of integrated recovery operations is based on the adsorption of proteins to particulate adsorbents in the presence of biomass. Usually, adsorbent particles are packed in fixed beds which have to be challenged with clarified feed solutions, because they act like filters in collecting all particulate material on top of the bed, which then is blocked and can no longer be used for protein adsorption. The solution to this problem is found in the increase of the interparticle distance of the adsorbent bed, thus allowing particles contained in the feed to pass freely in the interstitial volume of the matrix. Although there are several ways putting this concept into practice, fluidizing the adsorbent particles in a liquid flow directed upward seems to be the most widely distributed method [11]. By introducing biomass-containing feedstocks to a fluidized matrix chromatography becomes possible starting from unclarified biological suspensions, thus combining clarification, concentration, and initial fractionation in a single operation in the sense of integrated processing. Figure 1 shows, how a fluidized bed adsorption step can be integrated into a conventional purification protocol, thus reducing the number of initial recovery operations with the aim of a condensed, faster, and more efficient purification process with increased yield.

1.3 Scope

The present paper deals with fluidized bed adsorption as an integrative recovery operation. The scope of the contribution is first to describe the concept of the method and the different principles of achieving a combination of solid-liquid separation and chromatography. In the following the main system

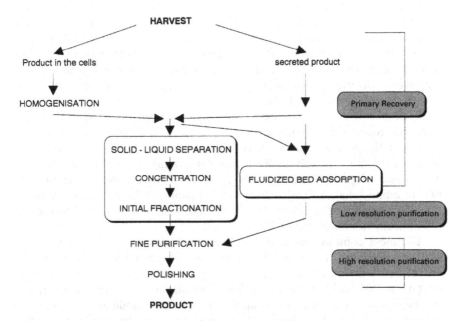

Fig. 1. Implementation of fluidized bed adsorption into a conventional procedure of protein purification

parameters governing the success of fluidized bed adsorption will be discussed and measurements as well as parameter estimations will be referenced to identify possible limiting steps. The paper will be completed by examples of successful implementations of the technology into protein purification protocols. In summary and conclusion, some perspectives, drawbacks, and possible pitfalls will be discussed in order to give a realistic outlook of what the potential of the method might be in the large scale recovery of proteins.

2 Principles of Fluidized Bed Adsorption

2.1 Methods of Whole Broth Processing

There are several possible arrangements tolerating the presence of particles during adsorption of proteins to particulate matrices. Batch adsorption in stirred tanks is performed by contacting adsorbent particles with a cell containing suspension. After protein capture the adsorbent is separated from the broth and the protein of interest can be eluted. This procedure has been described for the isolation of antibiotics [12], the purification of α-amylase from *B. amyloliquefaciens* broth [13], and the isolation of the prothrombin complex from cryosupernatant of human plasma [14] using commercially available resins for packed bed chromatography. The immobilization of conventional adsorbent particles in hydrogels was introduced to protect the matrix from fouling by cells or cell debris when contacted with crude feedstocks in a batch adsorption. Nigam et al. described this procedure at the example of β-lactamase purification from *B. stearothermophilus* [15].

Batch adsorption in stirred tanks is a one-stage process which usually has lower resolution than frontal chromatography, additionally the problem of separating the protein loaded matrix particles from the biomass subsequent to adsorption has to be solved. Real chromatography of unclarified broth on packed beds of specifically designed large beads has been shown for the purification of monoclonal antibodies from cell-containing hybridoma culture broth [16]. The specific surface chemistry of the large beads was described to prevent cell attachment and blocking of the packed bed.

The most common method of whole broth adsorption with particulate matrices is by fluidizing the particles. Development of a stable fluidized bed with an increased interstitial volume is the key to the purification of biomass-containing feedstocks in this set-up. Two possible modes of protein adsorption can be realized: The feed can be applied in a frontal mode as in packed bed chromatography or the column effluent can be recyled, leading to a situation comparable to a batch adsorption in a stirred tank. Frontal and single stage recirculating modes are illustrated in Fig. 2.

2.2 Liquid Fluidization of Adsorbent Particles

If we consider a bed of adsorbent particles in a chromatography column, for a stable fluidization, the forces of particle-fluid interaction must balance the particle weight. This condition is met, if the velocity of a liquid stream directed upwards exceeds a minimum value, which is called the minimum fluidization velocity U_{mf}. In this case, a steady state with an enlarged particle distance is achieved as long as the liquid velocity is not increased above the terminal settling velocity of the fluidized particle U_t, defining the point when the supporting forces exceed the particle weight and the adsorbent is elutriated from the column. The range of flow rates which can be used is dependent on the properties of the fluidized particles and of the fluidizing liquid. The falling velocity of a single particle U_t is described by Stokes law (Eq. 1) as being dependent from the particle diameter d_p, the density of the solid (ρ_p) and liquid (ρ_l) phase and the viscosity of the liquid phase η.

$$U_t = \frac{(\rho_p - \rho_l) \cdot d_p^2 \cdot g}{18 \cdot \eta} \tag{1}$$

Equation (1) may be used to estimate the order of magnitude of the terminal settling velocity of adsorbent particles. As in a fluidized bed the particles cannot

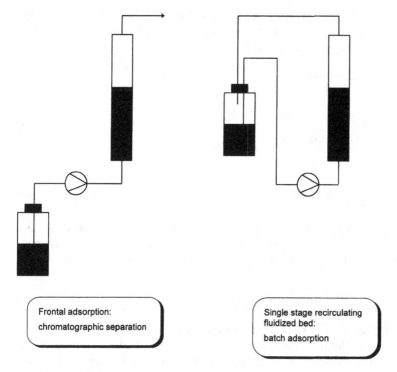

Fig. 2. Frontal and single stage recirculating fluidized bed adsorption

be treated as independent but interact due to adhesion forces, further considerations regarding fluidization of adsorbent particles in a column will be required.
There are numerous theories and correlations available in the literature describing liquid fluidization of particles, for a detailed analysis the reader is referred to
the excellent review by di Felice [17]. For our behalf the simple correlation of
Richardson and Zaki [18] will be sufficient. In this approach, the expansion of
a bed of particles in a liquid flow is described by Eq. (2), correlating the voidage
of the bed ε with a linear liquid velocity U by two parameters: the terminal
settling velocity of a single particle U_t and the expansion index n.

$$U = U_t \cdot \varepsilon^n \tag{2}$$

The two parameters are determined experimentally by measuring the bed void
volume at different linear velocities and a double-log plot of ε vs U. n and U_t are
found as slope and intercept of a linear regression of experimental data in this
plot. Additionally the parameters may be evaluated from literature correlations,
which give a good estimate of the range of flow rates applicable for the
fluidization of a certain matrix. Martin et al. [19] used two dimensionless

numbers for description of the fluidization behavior of particles, the Gallileo number Ga and the terminal Reynolds number Re_t, from which the terminal settling velocity and n are calculated.

$$Ga = \frac{\rho_p g(\rho_p - \rho_l)d_p^3}{\eta^2} \tag{3}$$

$$Re_t = \left[\frac{23}{Ga} + \frac{0.6}{Ga^{0.5}}\right]^{-1} \cdot \frac{1}{1 + 2.35\frac{d_p}{d_c}} \tag{4}$$

$$\frac{5.1 - n}{n - 2.4} = 0.016 \cdot Ga^{0.67} \tag{5}$$

The minimum Reynolds number for fluidization Re_{mf} as well as the minimum fluidization velocity U_{mf} may be calculated according to the correlation of Riba et al. [20], Mv defining the ratio of particle to liquid density $(\rho_p - \rho_l)/\rho_l$.

$$Re_{mf} = 1.54 \cdot 10^{-2} Ga^{0.66} Mv^{0.7} \tag{6}$$

This set of equations is sufficient to characterize a particulate matrix which should be used in fluidized bed adsorption regarding its fluidization behavior. It has to be noted, however, that the correlations have been developed for the fluidization of spherical particles of uniform diameter. In reality, most adsorbents are provided with a certain distribution of particle diameter. In this case, classified fluidization occurs and a modified equation should be used to describe the hydrodynamics of bed expansion [21]. For an estimation of the suitability of a certain matrix for fluidized bed adsorption the correlations shown above are convenient to use and provide sufficient information. The minimum fluidization velocity may be calculated using an average particle diameter as recommended by Couderc [22]. In the next section, conventional as well as new matrices shall be described under this respect.

In the equations shown above, the expansion of the fluidized bed at a given linear flow rate is determined by specific weight and diameter of the particles used as well as by the properties of the liquid phase. However, bed expansion can also be influenced by introducing mechanical stabilizers to the fluidized bed. Static mixers [23, 24] as well as dividing the column into different compartments by sieve plates [25] reduce bed expansion at a given linear flow rate and allow processing of liquids with increased viscosity without increasing particle diameter or density. Stable fluidized beds can also be developed employing an external magnetic field. By using particles, which are magnetically susceptible, the external field stabilizes the particles in a fluidized state independent of the flow rate of the fluidizing liquid [26]. In a certain range of linear velocity the strength of the external field then is the dominating parameter influencing the expansion of the bed.

2.3 Matrices for Fluidized Bed Adsorption

The correlations described above will now be used to check the suitability of matrices for protein chromatography as adsorbents in fluidized beds. In the catalogue discussed below, only those particles were included, which may be used to bind proteins in their three dimensional structure, e.g. porous adsorbents with sufficient pore diameter and protein compatible ligands attached to their internal surface. As a first example the physical data of Sepharose Fast-Flow (Pharmacia BioTech, Uppsala, Sweden) as a well characterized medium for packed bed protein chromatography will be discussed here. As can be seen from the terminal settling velocity denoted in Table 1, even for fluidization in water, the linear flow rate above which the matrix is driven out of a chromatography column is only 3.7 cm/min. If we now consider that fermentation broths or cell homogenates may have a significantly higher viscosity than water (e.g. 0.003 kg/(m·s) for a 3% wet weight yeast homogenate), then it is clear that increased particle density and diameter are necessary if matrices are to be used in fluidized bed adsorption. Chase and collaborators have determined a variety of data employing standard Sepharose adsorbents in fluidized bed adsorption [27–31]. Besides many other conclusions from their pioneering work, the authors demanded a suitable combination of particle diameter and density for a sufficient difference in terminal settling velocity between adsorbent and biomass contained in the fluidizing liquid [30]. The basic requirements formulated in this work provide a most useful set of criteria for choosing a matrix for fluidized bed adsorption. Dasari et al. presented a set of data on conventional glass-based adsorbents, characterizing the fluidization by experimentally determined terminal settling velocities and Richardson-Zaki coefficients [32]. From the matrices described only Fractosil 1000 with an apparent density of 1389 kg/m^3 had suitable properties. The use of controlled pore glass in fluidized bed adsorption was described by Thömmes et al. [33], showing that terminal settling velocities of 29 cm/min make this material an interesting candidate. A commercially available Protein A affinity adsorbent based on controlled pore glass has also been characterized [34], analyzing the data presented according to Richardson and Zaki yields an U_t of 9 cm/min and a fluidization index n of 5.9

Table 1. Estimation of fluidization of Sepharose FastFlow

Particle diameter (m)	$9 \cdot 10^{-5}$
Particle density (kg/m^3)	1131 (29)
Liquid density (kg/m^3)	1000
Liquid viscosity (kg/m·s)	0.001
Column diameter (m)	0.05
Ga	1.33
Re_t	0.056
U_t (cm/min)	3.7
n	5.05
U_{mf} (cm/min)	0.34

Gilchrist et al. presented a composite material from titanium dioxide and cellulose with 1200 kg/m³ density and compared it to glass-based matrices such as Spherosil and Spherodex [35]. The fluidization behavior of all materials was shown to be satisfactory, demonstrating, that a variety of matrices suitable for protein chromatography may be fluidized in a reasonable range of linear velocities. Agarose-based materials have been commercialized especially for fluidized bed adsorption by increasing their specific weight with incorporated quartz particles. There are several studies on fluidization of the so-called Streamline materials available and they all show a useful range of linear velocity from 2–10 cm/min [36–38]. An alternative method of increasing the specific weight of agarose particles is the incorporation of stainless steel particles. Fluidization of this second generation of Streamline adsorbents was presented by Thömmes et al. [39]. In this case the increased density of the steel weighted adsorbents allowed the use of smaller particles, which is beneficial to protein adsorption in a fluidized bed due to improved mass transport capabilities, as will be shown later. A very interesting matrix for fluidized bed adsorption is zirconia. As this material has a significantly higher specific weight than silica, small particles may be fluidized at linear flow rates similar to those used for silica or weighted agarose particles having a much higher diameter as indicated by Morris et al. [40]. Perfluoropolymer particles of 2200 kg/m³ density develop stable fluidized beds which may be used for protein purification when the support is derivatized with suitable ligands. The increased density of the support also allows the use of comparatively small particles (d_p 50–80 μm) at reasonable linear flow rate (2 cm/min) [41, 42]. Chetty and Burns have shown, that non-magnetic particles may be stabilized in a magnetic field, if they are supported by a fluidized bed of magnetized nickel particles [43]. The magnetic field acting on the nickel particles causes interparticle forces to influence the fluidization of the non-magnetic particles as well. This multiple-support magnetically stabilized fluidized bed (MSMSFB) may then be used for protein adsorption.

The adsorbent particles which have been described as suitable in the literature so far are summarized together with their fluidization parameters in Table 2. In this table the terminal settling velocity and the index n according to the Richardson and Zaki plot were taken from the literature when available, in some cases indicated by an asterisk plots of bed height vs flow rate available in the literature were evaluated according to Richardson and Zaki. If no such data were provided, the minimum fluidization velocity U_{mf} as well as n and U_t were calculated according to Eqs. (4–6) respectively. In cases when the adsorbents did not have a single particle diameter but a distribution in particle size the mean diameter was used for the calculations. Additionally the range of linear velocities used for fluidization of the material as described in the respective papers is shown. To make the data comparable, only values measured with water or aqueous buffer as the fluidizing medium were included in the table, calculated parameters were obtained with viscosity and density of water ($1 \cdot 10^{-3}$ kg/(m·s), 1000 kg/m³).

Table 2. Fluidization characteristics of different matrices suitable for protein adsorption

Matrix	U_t (cm/min)	n	U_{mf} (cm/min)	Range of linear velocity (cm/min)	Ref.
Lewatit S100	90	n.a	4.8	6–60 (tube) 12–60 (plate column)	[25]
Macrosorb K6AX	111[c]	4[c]	3.7[c]	24–90	[53]
Q Sepharose Fast Flow		4.7	0.04	0.1–0.7	[28]
Spherosil QMA	48.8[c]	4.6[c]	2.9[c]	3–14.4	[23]
DEAE Spherodex LS	41.4	3.9	0.7	1.2–18	[85]
Fractosil 1000	15.5*	3.9*	1.2[c]	1.2–6.5	[32]
Prosep A	8.9*	5.9*	n.a.	0.5–5	[34]
Prosep-5CHO	4.2*[a]	5.4*[a]	0.5[c a]	0.75–2.25	[48]
Ti-Cell-1	60[c]	4.4[c]	1.9[c]	0.9–18	[35]
DEAE Spherosil	22[c]	4.9[c]	1.7[c]	0.9–16.2	[35]
DEAE Spherodex	24[c]	4.9[c]	1.8[c]	0.9–16.2	[35]
Streamline	10.2–21.6**	3.1–4.8**	0.24–0.6**	0–7.2	[36]
Perfluoropolymer particles	n.a.	n.a.	n.a	1.4–2.1	[41]
Streamline quartz weighted	23.4*	5.4*	n.a.	1.7–8.3	[51]
Bioran CPG	28.8	4.5	1.4[c]	1.4–8.2	[33]
Streamline Red H-E7B	14.8	4.8	n.a.	0.3–4.2	[38]
Streamline SP	26.4	4.7	0.3	0.8–5	[37]
Streamline rProteinA	16.6	3.4	1.9[c]	1–7	[39]

[c] calculated according to Eqs. (4–6), physical parameters of the liquid set to 1000 kg/m^3 liquid density and $1 \cdot 10^{-3}$ kg/(m·s) liquid viscosity
* data available as bed height vs. flow rate were plotted according to Richardson and Zaki and parameters were evaluated after linear regression
[a] Data were obtained using cell free mammalian cell culture supernatant as liquid medium, viscosity and density comparable to aqueous buffer solutions
** Different values were obtained depending on liquid distribution at the column inlet

2.4 Classified Fluidization

In a fluidized bed, there is a principal difference in the overall mixing behavior compared to a standard fixed bed used for protein chromatography: the mobility of the particles adds the dispersive element of solids mixing which compromises the efficiency of the sorption process [11]. As mentioned above, adsorbents for protein chromatography usually do not have a single particle diameter but have a range of particle sizes. Liquid fluidized beds of these matrices show classification, a phenomenon which can simply be described as sorting out of particles by size, where large particles are found at the bottom of the bed and small particles are distributed towards the top of the column. This segregation behavior restricts the local mobility of the fluidized particles, so in extreme cases, the bed may be considered as consisting of layers of particles which are strictly separated and do not mix. This bed has dispersion characteristics similar to a packed bed and is supposed to yield comparable sorption efficiency. The degree of classification is dependent on the ratio of the size of the largest and smallest particle within the bed. Al Dibouni and Garside [21] define a size ratio > 2.2 as sufficient for the formation of a "perfectly classified bed". The specific details of solids and liquid mixing and the consequences to the adsorption process will be discussed below. At this point, it is sufficient to note that a classified fluidization stabilizes the adsorbent bed by a reduction in the mixing of solids. Whereas, in most of the adsorbents described in Table 2, the particle size distribution usually present in batches of these matrices is more or less "accidental", in new commercial preparations of matrices for fluidized bed adsorption the classified fluidization is effected deliberately by producing adsorbents with a defined distribution in particle size and/or density (particle size ratio of 2.5). To account for the difference in the dispersion characteristics of the classified bed from a fluidized bed of monosized particles, the fluidized beds from adsorbents with the size and/or density distribution have been called expanded beds by several authors [11, 37, 38, 44–47]. The classified fluidized bed still fulfils the basic condition for complete fluidization, a linear velocity of the fluidizing liquid which is higher than the minimum fluidizing velocity of the matrix used, so the expanded bed for protein adsorption may be considered as a special case of a simple fluidized bed.

2.5 Column Design for Fluidized Bed Adsorption

The development of a stable fluidized bed is not just dependent on the stationary phase, the column used also has to fulfil some simple but very important demands. A suitable liquid distribution is crucial for bed stability, especially if a classified fluidization is necessary for a successful sorption process. The basic demand is an evenly distributed pressure drop across the distributor plate at the column inlet thus supporting a flat velocity profile in a cross section of the column. Pressure drop fluctuations lead to the development of channels of preferential passage of liquid, the most important inhomogeneity in an adsorption process. Several plate configurations have been tested and their influence on matrix expansion and bed stability has been investigated [36]. Alternatively a bed of glass ballotini may be used as a liquid distribution system [34, 39, 48]. Bascoul et al. provided investigations on bed stability depending on distributor design showing, that channeling in the lower part of a fluidized bed due to uneven flow distribution is reduced with increasing column length. This observation induced the conclusion that a fluidized bed is the best distribution system for a fluidized bed [49]. As fluidized beds of adsorbent particles are supposed to allow whole broth processing the distributor has to enable the passage of biomass without being blocked and without damaging shear sensitive cells. Cell breakage in the flow distributor leads to the release of intracellular compounds and may severely reduce the success of the protein purification. Another very important parameter influencing bed stability is the alignment of the column. Van der Meer et al. have shown, that very small deviations from vertical alignments lead to significant inhomogeneity of liquid flow and to reduced bed stability [50]. Therefore the importance of the careful experimental set-up of a column used in fluidized bed adsorption has to be emphasized.

As long as these basic design rules, even pressure drop across the column inlet, allowing undamaged passage of biomass, and vertical alignment are taken into consideration, standard columns used for packed bed adsorption may be used. Additionally, specifically designed columns for fluidized bed adsorption are available on the market, where these aspects have been given special attention.

2.6 Experimental Procedure

In principle, the experimental protocol of fluidized bed adsorption does not deviate from packed bed operations, the main difference being the direction of liquid flow. The standard sequence of frontal chromatography, equilibration, sample application, wash, elution, and cleaning (CIP) is performed with an upward direction of flow as shown in Fig. 3. During equilibration of the matrix the stabilization of the fluidized bed occurs, in case of size and/or density distribution of the adsorbent particles the classification within the bed may be detected by visual observation of the bed. As discussed below, bed stability may

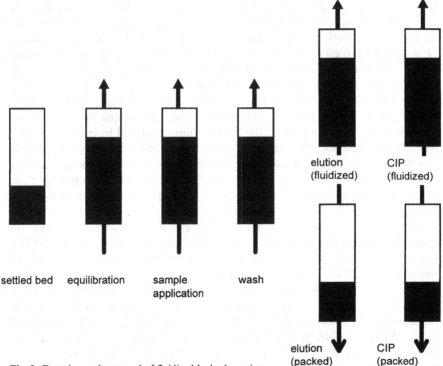

Fig. 3. Experimental protocol of fluidized bed adsorption

also be detected from a residence time distribution (RTD) in the liquid phase. On a large scale, when visual observation is not possible in large stainless steel columns, determination of the RTD of a suitable tracer pulse may serve as an indication of the development of a stable bed. Subsequently the biomass-containing broth is applied, after sample application, residual biomass and unbound proteins are removed from the bed in a washing procedure. Elution may be performed in two ways: as the washing procedure is supposed to remove all residual particles, the bed may be allowed to settle and elution can be performed in a packed bed mode with a reversed flow direction. For this purpose the top adapter, which is adjusted above the surface of the expanded bed during equilibration, sample application, and wash, has to be lowered to the top of the settled bed. This requires a movable adapter which is sealed against the environment by O-rings. In the case of pharmaceutical production, movable parts may be unwanted due to biological safety reasons, especially if production in a closed system is necessary. Therefore elution may be performed in a fluidized mode, the only difference being an increased elution volume due to the larger interstitial volume of the fluidized bed [39, 51]. With the advantage of being able to do without the movable adapter, this drawback may be acceptable.

The washing and cleaning procedures are very important in fluidized bed adsorption, as working with particle-loaded whole broths significantly increases the physical challenge of the matrix. Removal of residual particles may be tedious and require large washing volumes, thus increasing process times and medium cost. Hjorth et al. [51] demonstrated, that 20 sedimented bed volumes of washing solution were necessary to reduce the cell load by 10^5, when a fluidized bed of Streamline adsorbents was challenged with an E. coli homogenate containing 10^9 colony-forming units per ml. Similar results were obtained by Hansson et al. with whole E. coli cells [46]. Draeger and Chase [29] as well as Chang et al. [38, 44] have demonstrated, that increasing the viscosity of the washing solution reduces the liquid volume required to remove biomass from the fluidized bed. The clarification efficiency of the integrative procedure is directly correlated to the efficiency of the washing procedure. Residual particles still bound to the matrix may be removed during elution and contaminate the eluate thus necessitating another clarification step prior to further purification of the protein of interest. In the case of E. coli homogenate, reduction of particle concentration by a factor of 1000 is possible in an industrial centrifuge. As shown by Hjorth et al. this may easily be achieved using an expanded bed [51]. During adsorption of a recombinant fusion protein secreted by E. coli to fluidized Streamline DEAE the optical density at 600 nm representing the concentration of cells was reduced 250 fold, denoting efficient clarification [46]. For a mammalian cell system the particle load in the eluate of a fluidized bed purification of IgG2a was reduced more than 200 fold [39]. When a culture broth of the yeast Hansenula polymorpha was applied to a fluidized cation exchange adsorbent for the capture of recombinant aprotinin, the eluate could be transferred directly to HPLC purification without intermediate treatment, thus demonstrating the integration of clarification and capture during fluidized bed adsorption [52].

Application of whole broth to adsorbents for protein chromatography increases the contact of the matrix with nucleic acids, lipids, and cellular compounds which are removed in conventional primary recovery steps prior to standard column chromatography. Therefore specific care has to be taken to achieve a thorough cleaning of the adsorbent. This is especially valid for adsorption of particles e.g. cell debris to the adsorbent surface, which can induce aggregation of adsorbent particles within the bed during the next sample application. Aggregates of particles have different terminal settling velocities due to their increased diameter, lead to the formation of flow channels within the bed, and hamper the formation of the perfectly classified fluidized bed with negative consequences for the efficiency of the sorption process. This aspect was discussed by Chang and Chase who provided a cleaning-in-place procedure based on NaOH, NaCl, and ethanol [44] as well as 4M urea [38].

3 Parameters Influencing the Performance of Fluidized Bed Adsorption

3.1 General Remarks

The performance of a conventional method of protein purification may generally be characterized by its capacity and resolution, in the case of a primary recovery operation the additional criteria of clarification efficiency and reduction of process volume become important parameters describing a successful procedure. In packed bed chromatography of proteins on porous adsorbents there are four main system parameters influencing the overall performance:

- binding equilibrium
- dispersion in the liquid phase
- fluid side mass transport
- particle side mass transport.

As in fluidized bed adsorption, proteins are bound to porous particles as well, these parameters will remain important and must be considered when describing protein adsorption to fluidized beds. As mentioned above, fluidizing the adsorbent allows free movement of the particles within the adsorbent bed, so dispersion in the solid phase is another component determining process performance.

The scope of the following chapter will be an attempt to identify system parameters limiting the efficiency of fluidized bed adsorption and to define operating parameters, e.g. adsorbent design, flow rate, column geometry, and

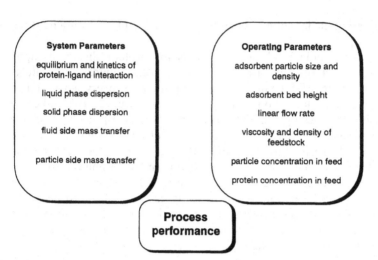

Fig. 4. System and operating parameters influencing process performance of fluidized bed adsorption

feedstock properties suitable for controlling this limitation. Defining suitable resins for this technology not only relies on the estimation of the fluidization properties of the material, a detailed understanding of the influence of particle size, density, and porosity on protein uptake in a fluidized bed is inevitable for a directed design of new fluidized bed adsorbents. The same is valid for process design with respect to the operating conditions of whole broth chromatography. The influence of the feedstock properties, linear flow rate of application, column dimensions etc. has to be described to allow a rational design. The interplay of different parameters is summarized in Fig. 4. In the following, the system parameters will be the focus of the discussion, the operating parameters will be included by describing their influence on equilibrium, hydrodynamics, and mass transport.

3.2 Sorption Equilibrium

As the ligand-protein interaction takes place at the internal surface of porous adsorbents, kinetics and equilibrium of the interaction should be independent of the interstitial voidage within an adsorbent bed. Therefore the equilibrium capacity of an adsorbent will not be influenced by different experimental configurations e.g. batch stirred tank, batch fluidized bed, frontal application to packed or fluidized beds. The major difference arises from the medium from which the protein is isolated. As fluidized beds are used for whole broth adsorption, the properties of the broth have to be considered regarding the possible influence of components which are removed in conventional primary recovery steps and therefore are not present during the initial chromatography operations in a standard downstream process. These are on one hand nucleic acids and lipids, which are often removed during concentration steps by ultrafiltration, on the other hand small particles such as cell debris or small bacterial cells may influence the binding capacity. Chase and collaborators have gathered data on the influence of different cell types on binding equilibrium of model proteins to conventional agarose particles. *Saccharomyces cerevisiae* cells at 3% dry weight reduced the equilibrium capacity of BSA binding to Q Sepharose by 49% [30]. The capacity of cation exchange [31] and affinity matrices [30] was only reduced by 13 and 17% respectively. Chase and Draeger additionally demonstrated the influence of different cell types on protein breakthrough in frontal adsorption [31]. From these experiments changes in equilibrium capacity as well as deteriorated mass transfer capability in the presence of cells may be detected. Very small concentrations of *Alcaligenes eutrophus* (0.5% dry mass) led to a fourfold reduction of Q-Sepharose capacity for BSA. Identical concentrations of yeast and *E. coli* cells affected capacity only slightly. The effects of different cell types on capacity were shown to be dependent on cell size and on the net charge of the cell surface, which is usually negative. From this point of view, the independence of the capacity of the cation exchange and affinity adsorbents from biomass may be understood. In addition to the reduced equilibrium capacity mass transfer to fluidized adsorbents was also hampered by the presence of cells, this will be discussed in detail below. In contrast to these findings, data reported by Wells et al. [53] demonstrate that the presence of *Candida olea* and *Saccharomyces cerevisiae* did not reduce equilibrium capacity of Macrosorb anion exchange resins during BSA adsorption in recirculating fluidized beds. This may be due to the different fluidizing conditions for the matrices: Macrosorb particles were fluidized at 36–42 cm/min linear flow rate, whereas the Sepharose adsorbents were fluidized between 0.6 and 1.6 cm/min. It has to be noted, however, that both materials were still fluidized in the laminar flow regime (Re_p 0.01–0.024 for Sepharose and 2.4–2.8 for Macrosorb). Gilchrist et al. [35] were able to show, that the presence of 2% dry mass of *S. cerevisiae* had an influence on capacity of several anion exchangers when BSA adsorption was performed in a single-stage recirculating fluidized bed (SSRFB). The reduction in capacity was found to vary between 20% for a matrix based on a titanium dioxide/cellulose composit (DEAE Ti-Cell-1) and 86% for the silica based DEAE Spherosil LS. During adsorption of glucose-6-phosphate dehydrogenase from unclarified yeast homogenates on commercially available anion exchange adsorbents for expanded bed adsorption (Streamline DEAE, application of 25% wet weight homogenate at 1.9 cm/min, $Re_p = 0.04$) Chang and Chase found no significant difference in the equilibrium binding capacity from clarified (17 U/ml) and unclarified (15 U/ml) feedstocks. Summarizing, the operating parameters influencing the binding equilibrium in packed beds e.g. protein concentration, conductivity, and pH are also valid in fluidized bed adsorption, additionally, the presence of particles, e.g. cells or cell debris, may reduce adsorbent capacity by attachment of cells to the adsorbent surface. In the case of ion exchange resins, this may be induced by electrostatic interaction between the charged protein surface and ion exchange ligands.

3.3 Dispersion in the Liquid Phase

The importance of liquid dispersion in chromatography has been proven extensively and there is no doubt that the distribution of residence time in a packed bed of porous adsorbents may severely affect the performance of the sorption process. A large amount of data is available on dispersion in packed beds and a variety of correlations have been proposed to describe the dependency of dispersion from linear flow rate in terms of particle Peclet number vs Reynolds number (for a short summary see e.g. [54]). Due to the mobility of particles, the fluidized bed is supposed to suffer from extended dispersion, therefore axial mixing of liquid elements has been the focus of several investigations in the field of fluidized bed adsorption. Early processes described for fluidized bed adsorption were performed in a well mixed or a recirculating mode. In these cases mixing behavior is close to that of batch adsorption in a stirred tank, yielding a one stage capture step. In cases where a frontal chromatography in a fluidized mode is desired, a separation process comparable to packed bed chromatography may be achieved, provided axial mixing is controlled properly.

Measurement of axial mixing in the liquid phase of a fluidized bed is performed by analysis of the residence time distribution of step or pulse signals [55]. By plotting the dimensionless E-function of the output signal versus the dimensionless time, the moments of the residence time distribution may be calculated according to Eqs. (7) and (8), the first dimensionless moment μ_1 describing the mean residence time and the second dimensionless moment μ_2 standing for the variance of the distribution.

$$\mu_1 = \frac{\int_0^\infty C t \, dt}{\int_0^\infty C \, dt} \tag{7}$$

$$\mu_2 = \frac{\int_0^\infty C(t - \mu_1)^2 \, dt}{\int_0^\infty C \, dt} \tag{8}$$

Characterizing the distribution according to the dispersion model yields a dimensionless number describing the degree of axial mixing within the bed. The Bodenstein number Bo relates convective transport of liquid to dispersion according to Eq. (9).

$$Bo = \frac{U \cdot L}{D_{axl}} \tag{9}$$

Here L is the length of the fluidized bed and D_{axl} is the coefficient of axial mixing in the liquid phase. In this case increasing column length results in an increased Bodenstein number, describing the reduced overall mixing in longer columns. Thus Bo is a measure of the state of liquid dispersion of a whole fluidized bed set-up and is best suited to describe a possible influence of axial mixing on the performance of a fluidized bed adsorption step. Characterizing mixing by

D_{axl} neglects the fact, that during passage through the bed the flow pattern is homogenized, underlining that D_{axl} may be higher at the bottom of the bed and axial mixing is reduced with increasing bed height as pointed out by Bascoul et al. [49]. Another dimensionless number for characterization of the liquid mixing is the particle Peclet number Pe_p (Eq. 10)

$$Pe_p = \frac{U_e \cdot d_p}{D_{axl}} \tag{10}$$

Here the particle diameter is used as the characteristic length and the influence of bed height on overall mixing is not accounted for.

When measuring axial mixing in the liquid phase of a fluidized bed using a tracer technique, the choice of the tracer may be a point to consider. As we are working with porous particles in protein adsorption, retardation of small tracer molecules in the pores of the adsorbent will occur. Comparison of mean residence time determined from the first dimensionless moment to the hydro-dynamic residence time calculated from bed voidage, flow rate, and column volume should reveal the degree of retardation. Additionally, tracers of reduced density compared to the density of the fluidizing medium may travel through the bed at increased velocity and therefore yield a residence time distribution which does not reflect reality. Tracer retardation during measurement of RTD was accounted for by Goto et al. [56]. In this paper an adsorption equilibrium constant K as well as a liquid to particle mass transfer coefficient k_f and a pore diffusion coefficient D_e were included into the description of the moments of the distribution according to Eqs. (11) and (12).

$$\mu_1 = \frac{L}{U}\left[1 + \frac{1-\varepsilon}{\varepsilon} \cdot K\right] + \frac{t_0}{2} \tag{11}$$

$$\mu_2 = \frac{2L}{U}\left[\frac{1-\varepsilon}{\varepsilon} \cdot \frac{K^2 r_p^2}{3} \cdot \left(\frac{1}{5D_e} + \frac{1}{k_f r_p}\right) + \frac{D_{axl}}{U^2}\left(1 + \frac{1-\varepsilon}{\varepsilon} \cdot K\right)^2\right] + \frac{t_0^2}{12} \tag{12}$$

Several tracers have been used in experiments describing axial mixing in fluidized beds of porous particles, e.g. acetone [37,57], Tryptophane [47], NaCl [49,56], radioactive tracers [58] and dextran blue [59]. It should be noted at this point, that measurement of RTD is not only important for determining possible domination of the chromatographic result by liquid mixing, Bo may as well be taken as a measure for the existence of a stable classified fluidized bed which is ready for sample application. Measurement of RTD in this case will provide a rational basis for the decision to start a large scale protein purification using a fluidized bed or to take measures for improvement of bed stability before application of valuable material.

Estimation of liquid mixing in a fluidized bed may be performed by a variety of correlations provided in literature. The problem in using the pro-posed correlations is, that they were obtained under many different operating conditions for a large number of different particles and are therefore only valid in a certain range of Reynolds numbers. The correlations will not be discussed in detail here, the reader is referred to the helpful introductions provided by Bascoul et al. [49] and by Slater [54]. In summary, most of the correlations predict increasing D_{axl} with increasing linear flow rate, the strength of the influence being dependent from the fluidized material. Probably the most interesting correlation for the case of protein adsorption was given by Tang and Fan [60]. For porous low density particles (ρ_p 1040–1300 kg/m^3) of larger size than usual in protein chromatography (d_p 1–2.5 mm) the authors proposed

$$Pe_p = 0.23 \cdot \left(\frac{\rho_p}{\rho_l}\right)^{-2.637} \cdot \varepsilon^{-2.4467} \qquad (13)$$

In Fig. 5, the liquid phase axial dispersion coefficient D_{axl} and the Bodenstein number Bo calculated from this relationship according to Eqs. (9), (10) and (13) are plotted for a range of linear velocities used in fluidized bed adsorption. The physical parameters of the commercial Streamline SP adsorbents (average particle size 247 µm, average particle density 1143 kg/m^3, terminal settling velocity 0.0044 m/s, $n = 4.7$ as described by Chang and Chase [37]) were

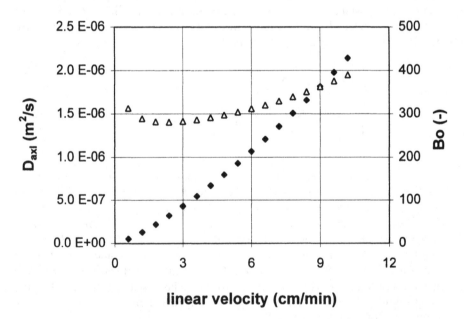

Fig. 5. Bodenstein number (Δ) and coefficient of axial dispersion (◆) versus linear velocity for a fluidized porous adsorbent calculated according to Eqs. (9), (10), and (13) [60], physical data of adsorbent: average particle size 274 µm, average particle density 1143 kg/m^3, terminal settling velocity $4.4 \cdot 10^{-3}$ m/s, $n = 4.7$ [37]

used for estimating the bed voidage according to Eq. (2). From this graph it may be seen, that D_{axl} in fact is increased with linear flow rate, but that the overall mixing situation, which is described by the Bodenstein number Bo (calculated according to Eq. (9)), follows a different pattern: with increasing linear flow rate axial mixing first is increased but due to the increasing column length during bed expansion the overall mixing will be reduced at higher flow rates. Although the values calculated deviate by a factor of ten from actual measurements (see below), the trend which has to be expected when fluidizing particles for protein adsorption is nicely predicted.

Experimental data on liquid dispersion are available for various fluidized matrices. Buijs and Wesselingh measured axial mixing of fluidized Lewatit ion exchange particles in a column consisting of different compartments which were separated by sieve plates [25]. Dispersion was described according to the tanks in series model, the tank number calculated from the moments of the distribution being 21, which is equivalent to a Bo of 42. The work presented by Chase and collaborators on fluidized Sepharose adsorbents also included RTD measurements [28]. The adsorbent particles, which had been originally designed for packed bed procedures, formed a stable classified fluidized bed with very low axial mixing in the liquid phase, dispersion coefficients were found to be $0.7\text{--}5.2 \cdot 10^{-8} \, m^2/s$ corresponding to Bodenstein numbers of 143. Here increasing linear flow rate also lead to higher D_{axl}, the overall mixing did not change as represented by constant Bo. Glass based porous adsorbents were investigated by Dasari et al. [32]. The authors described increasing D_{axl} with increasing linear flow rate, the overall mixing expressed as the column Peclet number analogous to the Bodenstein number showed no clear trend. Axial mixing was determined for two different particle sizes. Although indications were given, that fluidizing the larger particles results in less liquid dispersion, the differences were too small to allow for a definite judgement. Porous silica particles for enzyme immobilization were fluidized by Fauquex et al. [24, 58] employing static mixers for stabilization of the bed. Analysing the RTD according to the tanks in series model allowed the calculation of a number of stages (or tanks) per unit bed height. The number of stages initially decreased with linear flow rate and after passing a minimum rose again with increasing flow, analogous to the correlation of Tang and Fan as shown in Fig. 5. These findings were further supported by data from Thömmes et al. [59], where Bo followed a similar pattern. Additionally these authors demonstrated, that at constant linear flow rate overall mixing is reduced by increasing the column length, so matrices which do show increased liquid dispersion may be used nevertheless, if sufficiently long columns are employed. Comparing the absolute numbers describing liquid mixing with fluidized silica based adsorbents shows, that the flow pattern may well be described as plug flow in all these cases, so protein adsorption using these matrices under the conditions described presumably would not be controlled by liquid dispersion. This is either due to the very low linear velocities as in the case of the small particles investigated by Dasari et al. [32], to the use of static mixers stabilizing the fluidized bed [24, 58], or to a particle size distribution allowing

Table 3. Liquid mixing with fluidized silica particles for protein adsorption

Adsorbent	Particle size (μm)	Linear flow rate (cm/min)	Re_p	D_{axl} (m^2/s)	Bo	Ref.
Irregular Silica	100–160	6–21	0.03–0.5	$1 \cdot 10^{-4}$–$4 \cdot 10^{-6}$	15–90	[24, 58]
Lichroprep Si 60	25–40	0.2–1.5	0.001–0.009	$2 \cdot 10^{-7}$–$7 \cdot 10^{-6}$	7–20	[32]
Lichroprep Si 60	40–63	0.5–3	0.004–0.03	$8 \cdot 10^{-7}$–$8 \cdot 10^{-6}$	10–30	[32]
Bioran CPG	120–250	1–8	0.03–0.3	$4 \cdot 10^{-6}$–$6 \cdot 10^{-6}$	30–70	[59]

for the formation of a classified fluidized bed [59]. The values described are summarized in Table 3.

As discussed above, the formation of a classified fluidized bed is a condition for reduced mixing in a fluidized bed. Therefore weighted agarose adsorbents were made commercially available with a deliberately produced distribution of particle size and density. Axial mixing with these ion exchange matrices was described by several authors. All the data available showed, that liquid mixing was strongly reduced as was expected for a completely classified fluidized bed, the coefficients of axial mixing reported varying from 2–$9 \cdot 10^{-6}$ m^2/s [37,47,57]. Besides the effect of linear velocity, which was shown to increase D_{axl} at constant column Peclet number, Chang and Chase investigated the influence of feedstock viscosity on axial mixing [37]. They reported increasing D_{axl} and reduced column Peclet numbers if viscosity was increased at constant linear flow rate, an important finding, as crude feedstocks applied to fluidized adsorbents usually have increased viscosity depending on the biomass concentration in the broth. When the height of the fluidized bed was kept constant during viscosity increase by reducing the liquid flow rate, column Peclet numbers could be kept at the same level. This procedure was suggested as a possible solution for processing of highly viscous feedstocks without deviation from the plug flow type of liquid passage through the bed. Protein A coated agarose particles with specific weight increased by incorporated stainless steel spheres (average d_p 130 μm, mean ρ_p 1400 kg/m^3) were investigated by Thömmes et al. with regard to axial mixing [39]. Here a minimum linear velocity was described to be necessary for the development of a classified fluidized bed, overall mixing was then low corresponding to Bo of 33. A different method of reducing liquid dispersion is the application of an external magnetic field. The strongly reduced local mobility of the magnetized adsorbent particles allows a plug flow pattern of liquid passage through the bed [26]. Axial mixing in beds of magnetically stabilized Chitosan/Magnetite composites with a uniform particle size was compared to packed and non stabilized fluidized beds of these particles by Goto et al. [56]. In the MSFB D_{axl} was found comparable to the packed bed (0.5–$1 \cdot 10^{-6}$ m^2/s) but increased in the non stabilized fluidized bed (D_{axl} 1–$8 \cdot 10^{-6}$ m^2/s). From these data Bodenstein numbers of 25 may be calculated for the MSFB and of 6 for the fluidized bed. In this case protein adsorption to the fluidized matrix must be presumed to be dispersion controlled, a finding which is not astonishing considering the uniform particle size distribution which prevents the formation of the classified fluidized bed. From a dispersion point of view the MSFB will be well suited for protein adsorption, a prediction which has nicely been proven.

Summarizing the experimental data on liquid dispersion shows, that when porous particles suitable for protein adsorption are fluidized, a low degree of axial mixing in the liquid phase is found, if the fluidized bed is stabilized by reducing the local mobility of the fluidized particles. This may either be achieved by mechanical devices such as sieve plates or static mixers, the formation of a classified fluidized bed, or by applying an external field to magnetizable adsorbents. The condition of approximate plug flow has been fulfilled with each of these methods, therefore a possible limitation of the success of a fluidized bed adsorption process by enlarged axial mixing can be prevented by suitable stabilization of the fluidized bed. A minimum quality of fluidization is necessary for a successful protein adsorption to ensure liquid flow without preferential passage of liquid through channels existing in a poorly fluidized bed.

3.4 Dispersion in the Solid Phase

Movement of adsorbent particles within the bed as the most important difference between conventional fixed bed and fluidized bed adsorption is closely correlated to axial mixing in the liquid phase, therefore solid phase dispersion will be discussed as a system parameter in this section. Before describing experimental data, possible influences of solid phase dispersion on the sorption process shall be identified in a coarse picture. If we consider a protein solution moving through a fluidized bed under adsorbing conditions, then due to the plug flow movement of the protein solution, the bed may be described as slices which are subsequently filled with protein. If at a certain point, several particles already loaded with protein move in an upward direction into a zone which has not been reached by the protein front, then the adsorbed protein may be desorbed and washed out of the column with the protein depleted effluent. If this incident happens frequently, early protein breakthrough will occur. The same is valid for a situation when the protein front reaches the upper part of the bed. In this case, ideal plug flow behavior would allow adsorption of the protein front to the "unused" adsorbent slices. If mixing of particles occurs, then adsorbent particles already loaded will be distributed into the zone where protein adsorption is supposed to take place and reduce the local capacity in this area. Again early protein breakthrough will take place. As pointed out by Slater [54] these effects will not be important in the case of irreversible binding equilibrium when the sorbent capacity is independent from the protein concentration in the feed.

Up to now, there have been no data available in the literature describing solids mixing in fluidized beds of particles suited for protein adsorption. Therefore we will be restricted to a general discussion on particle movement and have to extrapolate some findings to fluidized bed adsorption. In fluidized beds of uniform particle size there are two possible ways of particle movement: the random motion in a perfectly mixed fluidized bed and the group-wise motion, which is induced by the strong viscous drag between the individual particles. As pointed out by Carlos and Richardson [61], the movement of the individual particle is random and may be described as a particle diffusion. There is, however, a significant contribution of convective particle movement to solids mixing, which is due to axial circulation, where particles rise in the center of the column and fall at the column wall (solid glass beads, d_p 8.8 mm). A similar pattern was described by Bascoul et al. [62], the velocity of the particle movement being increased with bed voidage. Column Peclet numbers based on particle dispersion were in the order of 1–2, indicating the high degree of mixing in the beds. For an efficient protein sorption which relies on plug flow movement of liquid and a stationary solid phase these conditions are unacceptable and reduce the chromatographic purification to a single stage capture comparable to batch adsorption in a stirred tank. The conditions are even worse in fact, as in the batch adsorption the residence time in principle is infinite, so mass transport limitations may be small, in a frontal adsorption residence time is low compared to the batch experiment, so in combination with severe mixing a successful adsorption process cannot be anticipated.

There are a few investigations on particle mixing in classified fluidized beds. Al Dibouni and Garside [21] described particle mixing in fluidized beds consisting of different size fractions of glass ballotini, showing, that the ratio of largest to smallest particle diameter determines the formation of the classified fluidized bed, a ratio of 2.2 being sufficient for perfect classification. The solids dispersion coefficient was described to be depending on bed voidage, an intermediate voidage of 0.7 resulting in a maximum dispersion. Similar results were shown by Kang et al. [63] and Yutani et al. [64] who also reported maximum dispersion at intermediate voidage for glass beads (d_p 1–6 mm, ρ_p 2500 kg/m³, narrow particle fractions [63]). In contrast to these data, van der Meer et al. [50] determined that D_{axp} increases steadily with linear flow rate using smaller and less dense particles (d_p 500–720 µm, ρ_p 1200–1400 kg/m³). Obviously the varying dependencies detected are an expression of the different experimental methods employed and individual measurements will be needed with porous adsorbents to allow an assignment to one of the relationships between U and D_{axp}. Van der Meer et al. as well as Kang et al. provided correlations for D_{axp}, which shall be used to estimate solids mixing with adsorbents for protein adsorption. The correlation of van der Meer et al. predicts

$$D_{axp} = 0.04 \cdot U^{1.8} \tag{14}$$

and from Kang et al. we obtain

$$D_{axp} = 2.97 \cdot 10^{-3} \cdot (U + U_{mf})^{0.802} \tag{15}$$

For controlled pore glass as it has been used for the adsorption of BSA and monoclonal antibodies in fluidized beds [59] (mean d_p 200 µm, ρ_p 1240 kg/m³) we find for 5 cm/min linear flow rate D_{axp} to be $1.2 \cdot 10^{-5}$ m²/s according to

Eq. (15) and $1.1 \cdot 10^{-7}$ m^2/s according to Eq. (14). This significant difference may be explained by the fact, the Kang et al. used particles of uniform size whereas van der Meer et al. measured dispersion between two fractions of different particle size in a segregated fluidized bed. As adsorption in a frontal mode is performed using classified or otherwise stabilized fluidized beds, the lower D_{axp} resulting from van der Meer's correlation may be a better description of the solid phase dispersion in a fluidized bed for protein adsorption.

As there are no experimental data on solids mixing with fluidized bed adsorbents, no definite statement on the role of solids dispersion as a system parameter in fluidized bed adsorption may be made. The calculation of D_{axp} for particles suited for protein adsorption according to van der Meer's correlation, which has been performed above, yields values of less than 10% of D_{axl} (for controlled pore glass $D_{axp} = 1.1 \cdot 10^{-7}$ m^2/s; $D_{axl} = 2-9 \cdot 10^{-6}$ m^2/s), so only a small influence of solids mixing on the performance of protein adsorption should be anticipated. Contrary to this estimation are experimental findings of Chang and Chase [37] during protein adsorption from high viscosity feedstocks. In this paper reduced sorption efficiency at increased solution viscosity was measured, which was attributed to increased liquid *and* solids mixing when the bed was expanded to a voidage of 0.9 at very high viscosity. As most feed solutions for fluidized bed adsorption, e.g. whole cell broths or homogenates, show increased viscosity, augmented solids mixing under these conditions may be limiting process performance. Therefore measurements of D_{axp} under defined operating conditions will be needed to allow a definite assessment of the role of solids dispersion as a system parameter in fluidized bed adsorption.

3.5 Particle Side Mass Transport

Due to the fact that protein adsorption in fluidized beds is accomplished by binding of macromolecules to the internal surface of porous particles, the primary mass transport limitations found in packed beds of porous matrices remain valid. Protein transport takes place from the bulk fluid to the outer adsorbent surface commonly described by a film diffusion model, and within the pores to the internal surface known as pore diffusion. The diffusion coefficient D of proteins may be estimated by the semi-empirical correlation of Polson [65] from the absolute temperature T, the solution viscosity η, and the molecular weight of the protein M_A as denoted in Eq. (16).

$$D = 9.4 \cdot 10^{-15} \frac{T}{\eta \cdot (M_A)^{1/3}} \tag{16}$$

Applying Eq. (16) to BSA the diffusion coefficient in free solution can be calculated as $7.3 \cdot 10^{-11}$ m^2/s, as shown by Skidmore et al. [66], however, this value was by a factor of 8.6 smaller when diffusion of BSA in a porous adsorbent was determined from batch protein uptake in a stirred tank. Diffusion of large molecules in the adsorbent pores is severely restricted, the extent of hindrance being increased with molecular weight, therefore pore diffusion frequently is the dominant transport resistance in conventional packed bed protein chromatography. As a consequence, the productivity of preparative protein purification using porous particles mostly is controlled by the residence time in the adsorbent bed [67]. Advanced adsorbents in protein chromatography aim at a reduction of this limitation by improved mass transport capability achieved by convective protein transport through perfusive pores [68, 69], accelerated diffusion in gels of high charge density [70], or by performing chromatography employing modified microfiltration membranes [71, 72]. As discussed above neither liquid nor solid dispersion seem likely to be the dominant system parameter as long as a minimum quality of fluidization is obtained and a restriction of particle motion is achieved, so there is no primary evidence, why fluidized bed adsorption should not also be a residence-time-controlled process.

The standard experiment to elucidate the limitation of protein adsorption by residence time is the measurement of protein breakthrough at varying linear flow rate. Plotting the dynamic capacity Q_{dyn} calculated from frontal protein adsorption versus the liquid residence time will show a decreasing Q_{dyn} as soon as pore diffusion is the dominant system parameter controlling the sorption process. Normalization of dynamic capacity to the equilibrium capacity Q_{max} calculated from the breakthrough curve is recommended, so the quotient of Q_{dyn}/Q_{max}, where both values are determined from the same breakthrough curve, will allow an unbiased view of the effectivity of sorption in a given experimental set up. These experiments have also been performed for fluidized bed adsorption. Hjorth et al. [51] measured dynamic capacity of lysozyme adsorption to a commercially available Streamline cation exchange matrix and of BSA to the respective anion exchanger. Increasing linear flow rate resulted in reduced dynamic capacity, the effect being more pronounced for the large protein BSA than for lysozyme. In terms of residence time controlled adsorption this was an expectable result, as the smaller protein is supposed to diffuse faster in the pores of the adsorbent. In a fluidized bed, it has to be noted, however, that increasing linear flow rate also leads to increased bed expansion and thus to increased total volume, a fact which has to be considered during calculation of residence time. The data on BSA adsorption provided by Hjorth et al. were recalculated in terms of Q_{dyn}/Q_{max} as well as in terms of residence time and the resulting plot is shown in Fig. 6. Due to bed expansion the residence time is not changed significantly if the linear flow rate is increased from 3 to 8 cm/min, nevertheless the dramatic decrease of Q_{dyn}/Q_{max} is found standing for a reduced sorption efficiency. A similar picture is found in the case of lysozyme sorption (data not shown). Additional insight may be found from data presented by Chang and Chase [37]. When increasing the solution viscosity by glycerol addition during measurement of dynamic capacity of lysozyme adsorption to Streamline SP a reduced dynamic capacity was found, although during viscosity increase the bed expansion and thus the liquid residence time was enlarged. Beyzhavi measured dynamic capacity of IgG1 binding to protein A coated controlled pore

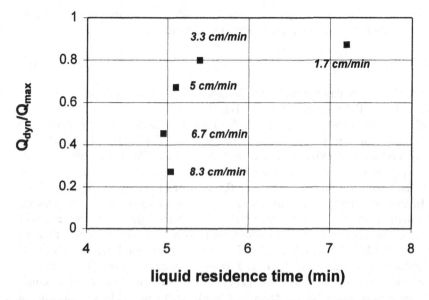

Fig. 6. Q_{dyn}/Q_{max} for BSA adsorption to fluidized Streamline DEAE at different linear flow rates. Original capacity data from Hjorth et al. [51], liquid residence time calculated from bed expansion data provided in Ref. [51]

Table 4. Adsorption of humanized IgG1 to fluidized Prosep-A. Capacity data taken from Beyzhavi [34]. Bed expansion was calculated according to Richardson and Zaki, with $U_t = 8.9$ cm/min and $n = 5.9$, the parameters were obtained as described in Table 2

Linear flow rate (cm/min)	Bed expansion (–)	Liquid residence time (min)	Dynamic capacity (mg/ml adsorbent)
0.5	1.6	19.3	40
1.7	2.5	11	38
2.5	3.1	10.1	37
5	6.5	11.8	34

glass (Prosep A) in a fluidized bed [34]. The results obtained with a varying linear flow rate are shown in Table 4, additionally the liquid residence time was calculated from bed expansion data provided in this paper. Again, increasing linear flow decreased the dynamic capacity although residence time was not reduced due to bed expansion.

Obviously liquid residence time is not an appropriate parameter to describe pore diffusion effects in fluidized bed adsorption. This may be elucidated by assessing particle side transport by a dimensionless analysis. Hall et al. [73] described pore diffusion during adsorption by a dimensionless transport number N_p according to Eq. (17), D_e denoting the effective pore diffusion coefficient in case of hindered transport in the adsorbent pores and U_e the effective linear flow rate (U/ε). Increasing N_p stands for improved mass transport capability and thus for enhanced sorption efficiency.

$$N_p = \frac{60 \cdot D_e \cdot (1 - \varepsilon) \cdot L}{d_p^2 \cdot U_e} \tag{17}$$

Here $L \cdot (1 - \varepsilon)$ is equivalent to the number of binding sites available in the adsorbent bed. If during fluidization L is increased at higher U, $(1 - \varepsilon)$ is reduced, which is consistent with the fact, that the amount of matrix in the bed, which is available for protein binding, is independent from the fluidization conditions. Thus increased bed expansion does not affect pore diffusion as expressed by N_p in spite of longer liquid residence time. The main influence on N_p is found from the effective diffusion coefficient D_e and from the particle diameter d_p.

In addition to linear flow rate and solution viscosity as operating parameters influencing particle side mass transport the biomass present in the feedstock during whole broth adsorption in fluidized beds has been investigated. Draeger and Chase [29] performed batch adsorption studies for binding of BSA to Q-Sepharose as well as of IgG to Protein A Sepharose in the presence of *Alcaligenes euthrophus* cells. The experimental data were used to calculate effective diffusivity of the proteins in the adsorbent pores. In the case of the affinity interaction no effect was found, whereas the diffusion coefficient of BSA in the pores of the ion exchange matrix was reduced 17 fold in the presence of 0.5% cells. Contrary to these findings Chase and Draeger [30] found only a 15% decrease in D_e when BSA was adsorbed to Q-Sepharose in the presence of 5% *Saccharomyces cerevisiae*. In this case fluid side mass transport was hampered severely represented by a 8 fold reduction in k_f. This discrepancy may be explained by the different size of the cells used in the studies. Analogous to the influence on equilibrium capacity, the small *A. eutrophus* cells may have entered the adsorbent pores and thus the mobility of the dissolved BSA molecules in the pores was restricted. The larger yeast cells may only have interacted with the outer surface of the adsorbent particles and thus the fluid side mass transport was obstructed. The affinity interaction was in no way blocked as there was no interaction of the cells with the Protein A ligands of the gel. These findings will be important, when cell homogenates are processed in fluidized bed adsorption: from whole cells no restriction in pore diffusion by the presence of cells will be expected, cell debris of very small size, however, may enter the pores and further reduce the effective diffusivity of the proteins leading to reduced sorption efficiency. In the investigations presented by Chang et al. [38] and Chang and Chase [44] on protein adsorption from yeast homogenates, reduced dynamic capacity in the presence of 25% homogenate was found for anion exchange and for a dye ligand modified material respectively. Possible reasons for this reduction were not discussed, however.

Summarizing the considerations on particle side mass transport, slow protein diffusion in porous adsorbents seems to have the same dominant influence on the efficiency of a fluidized bed adsorption as it frequently is the case in a packed bed. Contrary to conventional protein chromatography increasing linear flow rate leads to augmented bed expansion and thus to a relative invariance of liquid residence time in a certain range of flow rates. The fact, that above a certain liquid flow the dynamic capacity is reduced in spite of increased residence time suggests, that liquid residence time is not an appropriate parameter to describe the effect of pore diffusion in a fluidized bed.

3.6 Fluid Side Mass Transport

In advance to diffusive motion within a porous adsorbent to the binding sites situated on the internal surface the protein has to be transported from the bulk fluid to the outer surface of the adsorbent particle. For a conventional packed bed chromatography the transport coefficient k_f describing this mechanism has been correlated with dimensionless numbers as the Reynolds and the Schmidt group. For ion exchange of small ions, there are several correlations available, for protein adsorption in a fixed bed Skidmore et al. [66] used the equation of Foo and Rice (Eq. 18) to estimate k_f during modelling, the Schmidt number Sc being defined as $\eta/(\rho_l \cdot D)$.

$$k_f = \frac{D}{d_p} \cdot (2 + 1.45 \cdot Re_p^{1/2} \cdot Sc^{1/3}) \tag{18}$$

For transport of BSA ($D = 7.3 \cdot 10^{-11}$ m^2/s) in a packed bed of porous adsorbents ($d_p = 90$ µm) at 2 cm/min linear flow rate k_f is calculated as $9.4 \cdot 10^{-6}$ m/s when a voidage of the packed bed of 0.4 is assumed. As k_f is proportional to $Re^{1/2}$ fluid side mass transport efficiency is increased with linear flow rate and under conditions of standard packed bed protein chromatography this is not assumed to be the rate limiting step of the sorption process.

For the fluidized bed process the bed expansion as a consequence of an increase in linear flow rate has to be considered. In a simplified picture diffusive transport takes place in a boundary layer around the matrix particle which is frequently renewed, this frequency being dependent on velocity and voidage, as long as convective effects, e.g. the movement of particles are neglected. Rowe [74] has included these considerations into his correlation for k_f in fluidized beds, which is applicable for a wide range of Reynolds numbers, including the laminar flow regime where fluidized bed adsorption of proteins takes place (Eq. 19). The exponent m is set to 1 for a liquid fluidized bed, α represents the proportionality factor in the correlation for packed beds (Eq. 18) and is assumed as 1.45.

$$k_f = \frac{D}{d_p} \cdot \frac{2\xi/\varepsilon^m + \left[\dfrac{2\xi/\varepsilon^m(1-\varepsilon)^{1/2}}{(1-(1-\varepsilon)^{1/2})^2} - 2\right]\tanh(\xi/\varepsilon^m)}{\dfrac{\xi/\varepsilon^m}{1-(1-\varepsilon)^{1/2}} - \tanh(\xi/\varepsilon^m)}$$

$$\xi = \left[\frac{1}{(1-\varepsilon)^{1/2}} - 1\right] \cdot \frac{\alpha}{2} Sc^{1/3} Re^{1/2} \tag{19}$$

A different type of correlation which also includes bed expansion was presented by Fan et al. [75]. This equation was used by Chang and Chase to estimate k_f in protein adsorption to their expanded bed absorbents [37].

$$k_f = \frac{D}{d_p} \cdot [2 + (1.5 \cdot ((1 - \varepsilon) \cdot Re_p)^{1/2} \cdot Sc^{1/3})] \tag{20}$$

Both correlations are compared in Fig. 7 at the example of BSA adsorption to fluidized controlled pore glass (d_p 200 μm, ρ_p 1240 kg/m³, D 7.3·10⁻¹¹ m²/s, η 1·10⁻³ kg/(m·s). ρ_l 1000 kg/m³, bed expansion calculated according to Richardson and Zaki (Eq. 2), U_t and n estimated according to Eqs. 3–5). The same order of magnitude for k_f is predicted, Eq. (19), however, anticipates a steady decrease of k_f with voidage whereas Eq. (20) shows an optimum curve. The question, which correlation better describes reality can only be answered by measurements of k_f in fluidized adsorption systems, which are at the moment not available. Nevertheless we can easily see, that the simple assumption of increasingly effective fluid side mass transport at augmented linear flow rate does not hold for fluidized adsorbents. The absolute value of k_f in a fluidized bed will be lower than for a packed bed, an effect which should be considered during

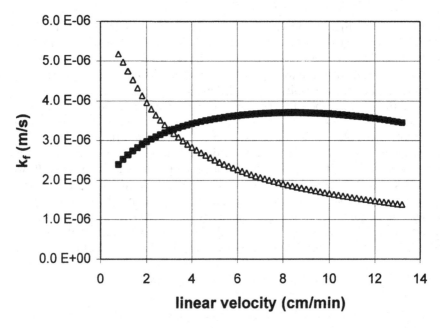

Fig. 7. k_f versus linear flow rate calculated for BSA adsorption to fluidized controlled pore glass after Rowe (△, Eq. 19, Ref. 74) and after Fan et al. (■, Eq. 20, Ref. 75). Physical data of adsorbent: average particle diameter 200 μm, average particle density 1240 kg/m³, BSA diffusion coefficient in solution 7.3·10⁻¹¹ m²/s, bed expansion calculated according to Richardson and Zaki, U_t and n estimated according to Eqs. (3–5)

protein adsorption. Chang and Chase [37] discussed fluid side mass transport to explain reduced capacity during expanded bed adsorption of lysozyme, which was found when protein sorption was conducted at enlarged solution viscosity and thus at higher bed voidage. No capacity reduction was found when the same experiments were performed in a packed bed, an additional hint, that the enlarged interstitial volume during bed expansion may influence the system parameter fluid side mass transport. Another supporting result is presented by Draeger and Chase [27]. Binding of BSA to fluidized Q-Sepharose was modelled based on equations set up for a packed bed adsorption and accounting for increased bed expansion. Comparison of the simulation and experimental findings showed, that the prediction of protein breakthrough was too optimistic in the early phase of the frontal adsorption. Reduction of the value of k_f, which had been estimated after Foo and Rice without allowing for an influence of ε, resulted in a better description of reality. As already discussed above, the presence of biomass had an influence on the value of k_f obtained in batch stirred tank experiments [29]. For fluidized bed adsorption in the presence of cells or cell debris, this may be an important operating parameter, if some influence on the sorption process in a fluidized bed is assigned to k_f.

The assessment of the role of k_f during protein adsorption in a fluidized bed may be performed with the help of a dimensionless transport number. Slater used the correlations provided by Rodrigues to simulate film transport limited adsorption of small ions to fluidized resins [54]. In this study dimensionless groups were used to describe the influence of the system parameters particle side transport, liquid dispersion, and fluid side transport. Dispersion was accounted for by the column Peclet number analogous to Bo as introduced above and mass transport from the bulk solution to the resin was summarized in a fluid side transport number N_L.

$$N_L = \frac{k_f \cdot a \cdot L}{U_e} \qquad (21)$$

Here a represents the surface area per column volume available for protein binding which is calculated from $a = 6 \cdot (1 - \varepsilon)/d_p$. If protein sorption is limited by fluid side mass transport, increasing N_L then stands for an enhanced transport efficiency represented by a steeper breakthrough curve in the initial phase and increased dynamic capacity during adsorption in a fluidized bed. In Fig. 8 the variation of N_L with linear flow rate is shown for the same system as in Fig. 7, k_f being estimated conservatively using Eq. (20). The major influence on N_L here comes from k_f, which may vary with changing operating parameters e.g. flow rate, feedstock viscosity, and particle size. As the product from L and a is independent from linear flow rate, bed expansion influences N_L via k_f. As may be taken from Fig. 8, the efficiency of fluid side mass transport is reduced with increasing flow rate, an effect which should be kept in mind during process design of fluidized bed adsorption.

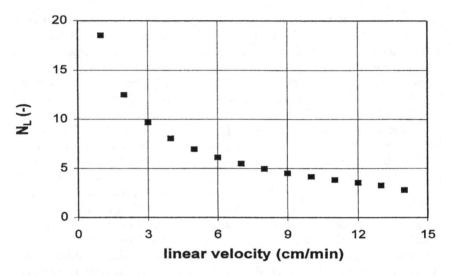

Fig. 8. Dimensionless fluid side transport number N_L (Eq. 21, Ref. 54) versus linear flow rate for BSA adsorption to fluidized controlled pore glass, for physical data of the adsorption system see Fig. 7

Summarizing this short discussion it has to be stated, that up to now experiments providing absolute numbers of k_f during protein adsorption in fluidized beds are not available, the interpretations are based on correlations derived for small ions. As ion exchange with fluidized resins is performed at much higher Reynolds numbers and mostly is not limited by particle side transport, the validity of the correlations for proteins has to be proven. Nevertheless, the influence of bed expansion at increased linear flow rate cannot be neglected and fluid side mass transport should be considered as a system parameter governing the sorption process in a fluidized bed under certain conditions.

3.7 Conclusions

After discussing the five system parameters which are possible candidates to be the rate limiting step during protein adsorption to porous particles in fluidized beds, the major impression may be that there are too many unknowns to allow a prediction, under which conditions such a process may be run successfully. It seems, that up to the present time no single parameter can be identified, which is solely controlling the efficiency of protein sorption in this operation, although particle side transport seems to be the dominant factor. The expansion of the adsorbent and the increase in voidage going with it causes a complex influence of operating parameters on the sorption process, as several system parameters are affected simultaneously. Veeraghavan and Fan [76] have developed a model for fluidized bed adsorption of organic acids to granular activated coal and compared their simulations to real experimental data. Under the conditions described mass transport to and within the adsorbent was limiting the process, it was stated, however, that different operating parameters will cause different limitations. In short columns, for example, liquid dispersion was assumed to be dominating the process, whereas in very long adsorbent beds solids dispersion was identified as a possible limiting system parameter. For the case of protein adsorption some simple calculations may help to elucidate the complex behavior. We choose a fluidized bed consisting of 600 ml of porous adsorbents (average d_p 200 µm, average ρ_p 1300 kg/m^3) which is supposed to capture a protein (D $7.8 \cdot 10^{-11}$ m^2/s) from aqueous buffer in a column of 5 cm diameter at 5 cm/min linear flow rate. Additionally we consider irreversible equilibrium, so capacity is equal for practically all protein concentrations, and dispersion in the solid phase shall be low compared to liquid mixing. The coefficient of axial mixing D_{axl} is assumed as $9 \cdot 10^{-6}$ m^2/s. All these parameters are realistic values for fluidized bed adsorption of BSA on porous glass or comparable adsorbents as Streamline, Spherodex, or Macrosorb. Fluidization is described according to Richardson and Zaki (Eq. 2), the terminal settling velocity U_t and n are estimated according to Eqs. (3–5). The residual system parameters liquid mixing, particle side and fluid side mass transport are described by dimensionless groups, liquid mixing by Bo (Eq. 9), fluid side transport by N_L (Eq. 21), and particle side transport by N_p (Eq. 17) representing a pore diffusion mechanism [73]. The effective diffusivity D_e is estimated as 20% of the diffusivity in free solution, k_f is calculated according to Fan et al. [75].

The respective coefficients as calculated according to these equations are found in Table 5. At twofold bed expansion, Bo of 56, N_L of 14.6, and N_p of 3.5 indicate a mass transport limited process. Lowering linear flow rate to 3 cm/min reduces Bo, but the enhanced mass transport expressed by higher N_p and N_L will certainly improve the sorption performance of the fluidized bed. If we choose smaller adsorbent particles (d_p 120 µm) of increased density (ρ_p 1500 kg/m^3) at 5 cm/min we see, that the reduced particle diameter leads to better mass transport capability and will result in enlarged dynamic capacity of the fluidized adsorbent compared to the larger particles. Assuming that reducing the linear flow rate to 1 cm/min will further improve performance we probably will find the opposite behavior: due to the reduction in Bo to 7.8 we produced a dispersion controlled process which no longer makes use of the improved mass transport properties of the new adsorbent. Although the exact values of this little

Table 5. System parameters calculated for different operating parameters

	d_p 200 m 5 cm/min	d_p 200 m 3 cm/min	d_p 120 m 5 cm/min	d_p 120 m 1 cm/min
Bed expansion	2	1.6	2.8	1.4
Bo	56	27	79	7.8
N_F	14.6	20.4	32.3	97.2
N_P	3.5	5.2	11.1	40.2

experiment may not meet reality in total, the complex interaction of system parameters due to the variable voidage is demonstrated. By varying the operating parameters, e.g. linear flow rate, adsorbent diameter and density, solution viscosity, and column length to diameter ratio the dominating influence of certain system parameter may be mediated or cancelled and another parameter may be controlling process performance.

3.8 Design Rules

Some basic design rules for fluidized bed adsorption may be deducted from the discussions on system parameters in this chapter. As mass transport is very likely to be slow due to the small diffusivity of protein macromolecules in solution, small adsorbent particles, where the distance, which has to be covered by the protein diffusively, is short should be favored. It has to be considered, however, that a reduction in particle size results in increased voidage at constant flow rate, which again deteriorates fluid side mass transport, so particle density has to be increased concomant to size reduction. Reduced dispersion in the solid phase, which will as well lead to lower axial mixing in the liquid phase, is achieved by classified fluidization, therefore a gradient in particle size and/or density in a batch of adsorbent will improve sorption efficiency. Alternatively magnetic stabilization or mechanical stabilizers, e.g. static mixers, will help to achieve this goal. Fluidized bed columns should be slender and long to allow for reduced liquid dispersion and sufficiently high mass transport capability within the bed. Column length is not limited by pressure drop as it is the case in a packed bed, so in principle there is no limit to optimizing mass transfer via column length. The improved overall mass transport in a long and slender column, however, is accompanied by a loss in productivity compared to a short bed of increased diameter, therefore a compromise has to be found between optimum matrix usage at high dynamic capacity in a long column and maximized throughput in shallow fluidized beds. A solution may be the subdivision of large diameter columns by vertical baffles, thus increasing the L/d_c ratio of short fluidized beds as suggested by Slater [54]. Choosing the correct process design of fluidized bed adsorption is a multidimensional optimization problem, which may be solved by careful control of the operating parameters and their influence on the system parameters governing the process.

4 Applications of Fluidized Bed Adsorption in Biotechnology

4.1 Antibiotics

Early work on the use of fluidized beds for whole broth processing in biotechnology was published in the field of isolation of antibiotics. In the paper published by Bartels et al. [77] the fundamental reason for investigations on integrated purification of biotechnologically produced molecules was presented: improvements of the performance of *Streptomyces* in streptomycin production by strain optimization resulted in changed physical properties of the culture broth, which led to large product losses during clarification in addition to reduced filtration flux. Therefore a process was developed which allowed direct contact of a cation exchange resin with the untreated fermentation broth. The tendency of the adsorbent particles to form aggregates in the broth necessitated the use of a mechanical agitator in front of the column outlet which was supposed to avoid blocking of the outlet screen. Comparing this new operation to the standard procedure of filtration and subsequent application of the filtrate to a packed bed showed a 12% increase in overall streptomycin yield. A similar procedure was published by Belter et al. on the recovery of novobiocin [78]. Large insolubles were initially removed on a vibrating screen and the resulting broth was then applied to a sequence of well-mixed fluidized beds of an anion-exchange resin. The operation was conducted in a semi-continuous mode with periodic elution and a 57% increase in overall yield was reported compared to a conventional sequence of filtration and packed bed chromatography. Large scale data on the purification of immunomycin by fluidized bed adsorption were provided by Gaillot et al. [79].

4.2 Proteins from Bacterial Cells

A very powerful tool for the expression of heterologous proteins is *E. coli*. In the following some examples of purification protocols for recombinant proteins produced in this organism are shown. The use of an expanded bed of Streamline DEAE to capture a secreted recombinant fusion protein was reported by Hansson et al. [46]. Fermentation broth containing 3.7% dry cell mass was diluted on line to reduce the ionic strength of the sample applied to the ion exchange process. The product was eluted at 90% yield, the process volume being reduced 16 fold. Additionally the particle content in the eluate as judged by OD at 600 nm was reduced 250 fold. In the eluate considerable amounts of DNA and endotoxin were found, further purification of the protein by immuno-affinity chromatography efficiently reduced the level of contamination, however. The purification of recombinant annexin V from *E. coli* homogenate was demonstrated by Barnfield-Frej et al. [45]. Small-scale packed-bed experiments for capture of the protein to Streamline DEAE were transferred to an expanded bed procedure employing 1.7 l of homogenate to 300 ml of adsorbent. Scale up to application of 26.5 l of homogenate of 4.7 l of matrix was demonstrated successfully at very high product yield (> 95%). The influence of biomass load in the feedstock on process performance was investigated: up to 5% dry weight was described to be applicable without problems, feedstocks containing higher biomass concentrations led to channeling within the bed and finally to a total collapse of the fluidized bed. Additionally feedstock viscosities of less than

0.01 kg/(m · s) were recommended as suited for this process. Spence et al. used activated controlled pore glass (Prosep-CHO) for immobilization of the inter-leukin 2 receptor [48]. The adsorbent was employed for fluidized-bed receptor-affinity chromatography of solubilized recombinant proteins from *E. coli*. Rec-ombinant human interleukin 2 and a single chain anti-Tac(Fv) *Pseudomonas* exotoxin fusion protein were refolded and the crude extract was purified in a fluidized bed procedure. The result of the procedures compared well to packed bed chromatography with clarified extracts.

4.3 Proteins from Yeast

Yeast is not only a very interesting natural source for enzymes as biocatalysts, it is also a very suitable host for the expression of complex recombinant proteins. Establishing protocols for fluidized bed adsorption of proteins from yeast is therefore an important step in validating this technology as a candidate for the large scale downstream processing of recombinant proteins. The first example was reported by Chase and Draeger [30] employing Cibacron Blue Sepharose FastFlow, an adsorbent initially designed for packed bed chromatography, in a fluidized bed for the isolation of phosphofructokinase (PFK) from a 5% homogenate of *Saccharomyces cerevisiae*. A 10-fold purified enzyme was ob-tained at 1.8-fold reduced process volume. A dye ligand coupled to an agarose-based matrix for expanded bed adsorption was described by Chang et al. during purification of glucose-6-phosphate-dehydrogenase (G6PDH) from baker's yeast homogenate [38]. The pseudoaffinity interaction of the enzyme with the Procion Red ligand resulted in a 103-fold purified enzyme at very high yield (98.8%). A similar procedure was described based on an anion exchange ex-panded bed adsorbent [44]. Starting from unclarified 25% baker's yeast homogenate a 12 fold purified enzyme was obtained at 98% yield. The enzyme was 2 fold diluted in the eluate, however. The suitability of perfluoro-polymer particles for fluidized bed adsorption was documented by Mc. Creath et al. [42]. Modification of small perfluoropolymer particles ($d_p = 50$–$80 \, \mu m$) with two different dye ligands (Procion Red H-E7B and Procion Yellow H-E3G) allowed the capture of G6PDH and malate dehydrogenase (MDH) from unclarified baker's yeast homogenate. The specificity of the dye ligand-dehydrogenase interaction was reflected by 103-fold (MDH) and 40-fold (G6PDH) purification. An approach to direct integration of a fluidized bed into a production process was shown by Morton and Lyddiatt. Fluidized anion exchange matrices were described as suited for the direct recovery of an acidic protease from unclarified broths of *Yarrowia lipolytica*. Purification in a fluidized bed allowed 2.5-fold faster processing of the enzyme compared to a conventional packed bed follow-ing microfiltration, resulting in a higher specific activity of the enzyme eluted after direct adsorption (9000 U/mg after packed bed process vs 13 000 U/mg after whole broth adsorption). This was attributed to less product autolysis in the shortened purification process [80]. By direct integration of a fluidized bed of Spherodex DEAE in an external loop of a batch fermentation of *Y. lipolytica* the enzyme was immediately removed from the fermentation broth and was thus protected from autolysis. Additionally synthesis of the growth related product was enhanced by the low stationary product concentration due to continuous enzyme adsorption [81]. An example for the direct adsorption of a recombinant protein secreted by yeast cells was given by Zurek et al. [52]. Recombinant aprotinin was purified from unclarified *Hansenula polymorpha* fermentation broth using fluidized Streamline SP adsorbent. The protein was recovered at 68% yield, the purification and concentration being 4 and 7 fold respectively. The clarified eluate could be further purified by HPLC without pre-treatment by filtration.

4.4 Proteins from Mammalian Cells

Examples for the purification of mammalian cell products employing fluidized adsorbents are up to now mainly restricted to monoclonal antibodies (MAb). Cation exchange procedures are well suited for the purification of MAbs from cell culture broth, as most of the antibodies have an isoelectric point (IP) between 5.5 and 8, whereas the main protein impurity in the culture BSA has an IP of 4.9. Therefore application at a pH between these two IP results in a discrimination of BSA binding and a substantial purification may be achieved. As mammalian cells are cultivated under isotonic conditions, capacity of cation exchange fluidized bed adsorbents is restricted by the high ionic strength in the medium. Additionally the reduction of pH to values below the IP of the antibody may damage the cells in the broth and lead to a release of intracellular material hampering the sorption process or resulting in secondary modification of the protein product. Experimental data on the adsorption of a recombinant antibody from CHO cell culture in a cation exchange process were provided by Batt et al. [47]. Volumes of 26 and 36 l of culture broth were allowed to settle overnight and the resulting supernatant containing $4.8 \cdot 10^4$ cells/ml was diluted threefold to reduce the ionic strength of the broth. Eight-fold purification and 13-fold concentration was achieved in the case of processing of a cell containing harvest after batch culture, whereas the same procedure resulted in 8-fold purification and 39-fold concentration when a harvest from continuous culture was used as feedstock due to the reduced initial concentration in the second case. The eluted fractions were described to contain some residual cells and cell debris. Calculation of process economics in comparison to a standard process of microfiltration/ultrafiltration followed by packed bed chromatography showed an overall cost advantage of the fluidized bed process at similar overall yield, the major disadvantage being the necessity of diluting the broth. A cation exchange procedure based on a fluidized controlled pore glass adsorbent as well as on a commercially available matrix (Streamline SP) was presented by Thömmes et al. [33]. In spite of increased liquid dispersion in fluidized beds employing the controlled pore glass compared to the agarose particles both adsorbents had similar efficiency in the purification of IgG2a from cell-containing broth using a 5 cm diameter column, thus underlining the mass transport controlled sorption of large protein molecules in a fluidized bed. An increase in scale of the process to a column with 20-cm diameter was described in a subsequent paper [82]. Born et al. reported the direct coupling of fluidized bed adsorption to a production process [83]. The effluent from a cultivation process using immobilized hybridoma cells in a fluidized bed bioreactor was directly applied to a fluidized bed of cation exchange adsorbents, the only manipulation being adjustion of pH to 5.5. The scope of this coupled process was the immediate reduction of the large liquid volumes, which usually have to be handled after continuous production employing high density cell culture bioreactors. Due to the whole broth adsorption using a fluidized bed, the biomass containing flow through fraction could be discarded without further handling. The MAb was obtained in a cell free eluate at fourfold purification and with a fourfold reduced process volume, thus simplifying further purification subsequent to the coupled process. In a paper by Erickson et al. [84] the isolation of a recombinant protein from CHO cells was shown. Operation was described on a pilot scale (d_c 20 cm) employing a Streamline SP cation exchange matrix. An important finding was, that levels of nucleic acids found in the eluate of the process were comparable to those found in eluates from a conventional packed bed process, where two subsequent filtration steps were needed for clarification. From these data it was concluded, that cells were not damaged during the process of sample application, although some concern still existed about the effect of the fluidized bed adsorption process on the quality of the eluted product for the reasons discussed above. In the same paper, a recombinant antibody was purified from unclarified CHO broth by fluidized bed adsorption to a protein A coated controlled-pore glass (Prosep A). The advantage of using affinity interactions in fluidized bed adsorption is, that capacity is not reduced at high ionic strength, so dilution is not necessary. Additionally the protein A-MAb affinity interaction may be exploited at pH 7, so cell damage due to pH adjustion is not anticipated. In the experiments described by Ericson et al. the capacity of the protein A matrix for the MAb was equal in packed and fluidized mode. As in the example with the Streamline adsorbent DNA content of the eluate did not deviate from measurements with packed beds. Problems were reported regarding the formation of aggregates from cells and adsorbent particles, the reason for the occurrence of the clogging was not discussed, however. Similar data were shown by Beyzavi [34] where IgG1 was purified from cell containing broth using the Prosep A material. Spence et al. demonstrated fluidized bed receptor affinity chromatography for the purification of humanized anti-Tac antibodies from whole culture broth [48]. An aldehyde activated controlled pore glass was used to immobilize the interleukin-2 receptor. The resulting affinity sorbent allowed the recovery of a fully active antibody after fluidized bed adsorption. An agarose based Protein A coated material for fluidized bed adsorption was described recently [39]. In these studies small scale experiments in a modified packed bed column were used as a basis for a scale up to a 5-cm diameter column. The adsorbent had a high capacity for the IgG2a tested allowing the purification of 2 g MAb in a single experiment employing 150 ml of matrix. Additionally data were provided on the clarification efficiency of the procedure, which demonstrated, that an initial retardation of particulate material (cells and cell debris) occurred during sample application. Experimental problems correlated to this retardation were discussed as well.

4.5 Miscellaneous Protein Sources

Natural protein sources in principle pose similar problems to the isolation process as cellular production systems. Protein extracts from animal or plant tissue or natural fluids usually contain particulate material, which has to be removed prior to chromatographic purification. Therefore fluidized bed adsorption can be considered as an alternative procedure for these feedstocks as well. Vouté described the use of fluidized DEAE Spherodex particles for the isolation of human serum albumin (HSA) from unclarified human plasma [85]. Three consecutive fluidized bed columns containing 10 kg adsorbent respectively were used resulting in 90% recovery of human albumin of 85% purity. The performance of the fluidized bed compared well with a packed bed process starting from pre-treated plasma. The major disadvantage of the whole broth process was, that the plasma had to be diluted 10.5 fold to achieve reasonable capacity. Therefore the HSA concentration in the eluate was 8 fold lower than after packed bed adsorption. The isolation of endo-polygalacturonase from a commercial enzyme preparation (Rapidase) was shown by Somers et al. [86]. Repeated batch adsorption of the enzyme to fluidized Ca-alginate could be operated for 100 cycles with consistent product output. Whey as a particle containing protein source was used during fluidized bed adsorption by Bascoul et al. [23]. α-Lactalbumin and β-Lactoglobulin were separated employing a silica based anion exchanger (Spherosil QMA) in a three stage fluidized bed process. Bed stability was increased by static mixers, thus allowing increased flow rates at constant bed expansion. The mechanically stabilized bed yielded higher productivity at increased flow rate compared to the process without static mixers. Scale up was demonstrated from a small column (d_c 5 cm) to a bed of 13 kg of adsorbent (d_c 40 cm). In their conclusions the authors stated the possibility of continuous processing of proteins in a fluidized bed in a counter current mode of operation. First realizations of this concept will be discussed in the next chapter.

4.6 Continuous Processing in Fluidized Beds

Continuous counter current operation of fluidized bed adsorption is used on a very large scale for small molecules, e.g. uranium or residues from mining. Theoretical and practical aspects of this technology have been summarized by

Slater [54]. For biotechnological products there are only few examples of counter current operation in fluidized beds. Burns and Graves described the use of magnetically stabilized fluidized beds (MSFB) for continuous processing [26]. Calcium alginate/magnetite particles were fluidized and the bed was stabilized with an external magnetic field. A continuous process was set up by a steady downward flow of adsorbent beads counter current to the steady upward flow of protein solution. HSA was bound from aqueous buffer to the Cibacron Blue modified magnetizable support in an initial process evaluation. Especially in counter current operation bed stabilization is important, as axial mixing affects the performance of the process. Conventional adsorbents may also be stabilized in a magnetic field, if a minimum amount of magnetisable particles is present as well in the fluidized bed [43]. This concept was verified for the continuous counter current separation of model proteins from aqueous solutions. A successful application of counter current adsorption of amino acids in a Cloete-Streat contactor was shown by Agosto et al. [87]. In this apparatus axial mixing was reduced by a compartimented set up, where the sections are separated by sieve plates. In their conclusions the authors stated, that this set up is in principle suited for the continuous processing of proteins from unclarified fermentation broths.

From these data we conclude, that continuous processing employing stabilized but non-classifying fluidized beds may be a valuable option for the large scale processing of bulk amounts of proteins. Practical examples of real protein products being isolated according to this concept are still lacking, however.

5 Concluding Remarks

As there are no universal concepts applicable to all possible problems in the primary recovery of proteins, a realistic judgement of the future role of fluidized beds for protein adsorption must include a discussion, under which circumstances fluidized beds will simplify the downstream process and where more problems are created than solved when this procedure is introduced to a protocol. As pointed out by Chase [11], there are situations where fluidized beds cannot be used due to specific properties of the biomass present in the broth. The major criterion for excluding whole broth processing is the occurrence of interactions between the adsorbent particles and the particulate material in the broth. A possible example is the formation of cell/adsorbent aggregates during sample application, leading to a deteriorated fluidization with channeling and dead zones in the bed with the well known consequences for the chromatographic performance. In the worst case total breakdown of the bed may happen. If these interactions cannot be eliminated by suitable flow conditions or additives, conventional solid-liquid separation by centrifugation or filtration is recommended prior to chromatographic purification. In case of very viscous feedstocks the terminal settling velocity of the adsorbent will be reduced to a degree, where the linear flow rates applicable are too low to allow for efficient separation between particulate contaminants and the matrix particles. In this case eluates may contain considerable amounts of particles and the demand of integration of chromatography and clarification is not fulfilled. Dilution of the broth can reduce this problem, but it has to be noted, that in this case the product of interest is diluted as well: a shift from favourable to unfavourable binding equilibrium may take place and system parameters, which have been neglectable using undiluted broth, e.g. liquid and solids dispersion, will be dominating the sorption process. Another important question is, how many manipulations of the broth are necessary to achieve conditions favouring the mechanism of the protein-matrix interaction. Some of these manipulations, e.g. high dilution to adjust ionic strength or adjustion to extreme pH values, will be damaging the biomass and lead to the release of additional components into the broth, which may hamper the sorption process or cause regulatory problems, e.g. in the case of DNA release from mammalian cells. The choice of the adsorbent should be adapted to these requirements.

If the principle suitability of the technology is ensured, then the design of the process will be determining if an advantage of integrative processing compared to the conventional clarification/concentration/packed bed adsorption approach is obtained. Choosing the correct operating parameters, e.g. bed height, linear flow rate, voidage etc., which effectuate protein adsorption will allow the establishing of a process of maximum performance. In this case, the combination of clarification, concentration, and initial purification using fluidized adsorbents will be a valuable alternative to a standard purification scheme.

Future developments are required in the quantitative description of system parameters. Neither solids dispersion nor fluid side transport have been quantified for the specific conditions of protein adsorption in fluidized beds: classified fluidization at very low linear flow rates is performed under conditions where most of the empirical correlations available from literature do not apply. Therefore experimental data under conditions suited for protein adsorption should be determined. Applications of this technology still are limited to comparably small scale, the demonstration of large scale applications is needed to validate the usefulness of fluidized bed adsorption as an integrated procedure. For continuous processing of very large amounts of protein fluidized bed adsorption in a counter current mode should be discussed, data showing whole broth processing in this kind of set-up will be a helpful extension of the existing body of knowledge. The number of protein-ligand interactions exploited up to now needs to be increased, especially robust pseudo-affinity interactions which are insensitive to the high ionic strength in fermentation broths may prove beneficial for use in whole broth processing. Further developments in matrix design should concentrate on the development of particles with enhanced mass transport capability, smaller beads with increased specific weight will be one possible route to improved sorption performance. The increased interest in academia as well as in industry to simplify downstream protocols for protein production has been beneficial to the development of integrated purification methods during the last few years, initial recovery of enhanced efficiency by combining clarification, concentration, and initial purification employing fluidized bed adsorption may be one possible contribution to achieve this goal.

Acknowledgements. This paper is dedicated to Professor Dr. Maria-Regina Kula, who has made invaluable contributions to the development of downstream processing of proteins for more than 20 years. The aspect of integrated processing of proteins employing methods as aqueous two-phase systems, membrane chromatography, and fluidized bed adsorption is found throughout her work and many of the developments in this field have been initiated and carried out by her vision of efficient protein recovery.

My thanks go to Almut Nagel, Klaus-Heinrich Gebauer, Holger Gieren, Jan Feuser, Andreas Karau, and Torsten Minuth for helpful discussions on integrated downstream processing in general and particularly on the details of fluidized bed adsorption as well as for critical reviewing of the manuscript.

6 References

1. Wheelwright SM (1989) J. Biotechnol. 11: 89
2. Datar RV, Cartwight T, Rosen C-G (1993) Bio/Technology 11: 349
3. Spalding BJ (1991) Bio/Technology 9: 229
4. Datar RV, Rosen C-G (1996) Cell and cell debris removal: centrifugation and crossflow filtration. In: Stephanopoulos G (ed) Bioprocessing. VCH, Weinheim, p 472
5. Kempken R, Preissmann A, Berthold W (1995) Biotechnol. Bioeng. 46: 132
6. van Reis R, Leonard LC, Hsu CC, Builder SE (1991) Biotechnol. Bioeng. 38: 413
7. Kroner KH, Schütte H, Hustedt H, Kula MR (1984) Proc. Biochem. 19: 67
8. Albertsson P-A (1960) Partition of cell particles and macromolecules. Almqvist and Wiksell, Stockholm
9. Kula MR (1990) Bioseparation 1: 181
10. Kroner KH, Krause S, Deckwer WD (1992) BIO forum 12/92: 455
11. Chase HA (1994) TIBTECH 12: 296
12. Krützfeldt R, Roß A, Deckwer WD (1992) BioEngineering 4: 32
13. Roe SD (1987) Whole broth extraction of enzymes from fermentation broths using commerically available adsorbents. In: Verall MS, Hudson MJ (ed) Separations for Biotechnology. Ellis Horwood, Chichester, p 210
14. Brummelhuis HGJ (1980) Preparation of the Prothrombin complex. In: Curling JM (ed) Methods of plasma protein fractionation. Academic Press, New York, p 117
15. Nigam SC, Sakooda A, Wang HY (1988) Biotechnol. Progr. 4: 166
16. Grandics P (1994) Direct purification of monoclonal antibody from hybridoma cell culture harvest on packed bed columns using novel large bead agarose ion exchange, affinity, and immobilized affinity media. Proceedings of the IBC meeting on monoclonal antibody purification, San Francisco
17. di Felice R (1995) Chem. Eng. Sci. 50: 1213
18. Richardson JF, Zaki WN (1954) Trans. I. Chem. E. 32: 35
19. Martin BLA, Kolar Z, Wesselingh JA (1981) Trans. I. Chem. E. 59: 100
20. Riba JP, Routie R, Couderc JP (1978) Can. J. Chem. Eng. 56: 26
21. Al-Dibouni MR, Garside J (1979) Trans. I. Chem. E. 57: 94
22. Couderc JP (1985) Incipient fluidization and particulate systems. In: Davidson JF, Clift R, Harrison D (ed) Fluidization. Academic Press, London, p 1

23. Bascoul A, Biscans B, Delmas H (1991) Rec. progr. gen. proc. 5: 275
24. Fauquex PF, Flaschel E, Renken A (1984) Chimia 38: 262
25. Buijs A, Wesslingh JA (1980) J. Chromatogr. 201: 319
26. Burns MA, Graves DJ (1985) Biotechnol. Progr. 1: 95
27. Draeger NM, Chase HA (1990) Modelling of protein adsorption in liquid fluidized beds. In: Pyle DL (ed) Separations for biotechnology 2. Elsevier, Dorking, p 325
28. Draeger NM, Chase HA (1990) I.CHEM.E. Symposium Series 118: 161
29. Draeger NM, Chase HA (1991) Bioseparation 2: 67
30. Chase HA, Draeger NM (1992) J. Chromatogr. 597: 129
31. Chase HA, Draeger NM (1992) Sep. Sci. Technol. 27: 2021
32. Dasari G, Prince I, Hearn MTW (1993) J. Chromatogr. 631: 115
33. Thömmes J, Halfar M, Lenz S, Kula MR (1995) Biotechnol. Bioeng. 45: 203
34. Beyzavi K (1994) Bioprocessing World 1: 5
35. Gilchrist GR, Burns MT, Lyddiatt A (1994) Solid phases for protein adsorption in liquid fluidized beds. In: Pyle DL (ed) Separations for biotechnology 3, The Royal Society of Chemistry, London, p. 186
36. de Luca L, Hellenbroich D, Titchener-Hooker NJ, Chase HA (1994) Bioseparation 4: 311
37. Chang YK, Chase HA (1996) Biotechnol. Bioeng. 49: 512
38. Chang YK, McCreath GE, Chase HA (1995) Biotechnol. Bioeng. 48: 355
39. Thömmes J, Bader A, Halfar M, Karau A, Kula MR (1996) J. Chromatogr. A 752: 111
40. Morris JE, Tolppi GC, Dunlap CJ, Carr PW, Flickinger MW (1994) Protein separation using HPLC and fluidized bed zirconia supports. Proceedings of the ACS Meeting, San Diego
41. McCreath GE, Chase HA, Lowe CR (1994) J. Chromatogr. A 659: 275
42. McCreath GE, Chase HA, Owen RO, Lowe CR (1995) Biotechnol. Bioeng. 48: 341
43. Chetty AS, Burns MA (1991) Biotechnol. Bioeng. 38: 963
44. Chang YK, Chase HA (1996) Biotechnol. Bioeng. 49: 204
45. Barnfield-Frej A-K, Hjorth R, Hammarstroem A (1994) Biotechnol. Bioeng. 44: 922
46. Hansson M, Stahl S, Hjorth R, Uhlen M, Moks T (1994) Bio/Technology 12: 285
47. Batt BC, Yabannavar VM, Singh V (1995) Bioseparation 5: 41
48. Spence C, Schaffer CA, Kessler S, Bailon P (1994) Biomed. Chromatogr. 8: 236
49. Bascoul A, Delmas H, Couderc JP (1988) Chem. Eng. J. 37: 11
50. Van der Meer AP, Blanchard CMRJP, Wesselingh JA (1984) Chem. Eng. Res. Des. 62: 214
51. Hjorth R, Kämpe S, Carlsson M (1995) Bioseparation 5: 217
52. Zurek H, Kubis E, Keup P, Hörlein D, Beunink J, Thömmes J, Kula MR, Hollenberg CP, Gellissen G (1996) Proc. Biochem. 31: 679
53. Wells CM, Lyddiatt A, Patel K (1987) Liquid fluidized bed adsorption in biochemical recovery from biological suspensions. In: Verall MS, Hudson MJ (ed) Separations for biotechnology. Ellis Horwood, Chichester, p 217
54. Slater MJ (1992) Principles of ion exchange technology. Butterworth and Heinemann, Oxford
55. Levenspiel O (1972) Chemical Reaction Engineering. Wiley, New York
56. Goto M, Imamura T, Hirose T (1995) J. Chromatogr. A 690: 1
57. Lindgren A, Johannson S, Nyström L-E (1993) Scale-up of expanded bed adsorption. In: Henon B (ed) BED - The seventh Bioprocess Engineering Symposium. The American Society of Mechanical Engineers, p 27
58. Fauquex PF, Flaschel E, Renken A, Do HP, Friedli C, Lerch P (1983) Int. J. Radiat. Isot. 34: 1465
59. Thömmes J, Weiher M, Karau A, Kula MR (1995) Biotechnol. Bioeng. 48: 367
60. Tang WT, Fan L-S (1990) Chem. Eng. Sci. 45: 543
61. Carlos CR, Richardson JF (1968) Chem. Eng. Sci. 23: 825
62. Bascoul A, Couderc JP, Delmas H (1993) Chem. Eng. J. 51: 135
63. Kang Y, Nah JB, Min BT (1990) Chem. Eng. Comm. 97: 197
64. Yutani N, Ototake N, Too JR, Fan LT (1982) Chem. Eng. Sci. 37: 1079
65. Polson A (1950) J. Phys. Colloid Chem. 54: 649
66. Skidmore G, Horstmann BJ, Chase HA (1990) J. Chromatogr. 498: 113
67. Janson JC, Peterson T (1993) Large scale chromatography of proteins. In: Ganetsos G, Barker PE (ed) Preparative and production scale chromatography. Marcel Dekker, New York, p 559
68. Afeyan NB, Fulton SF, Mazsaroff I, Regnier FE (1990) Bio/Technology 8: 203
69. Gustavsson P-E, Larsson P-O (1996) J. Chromatogr. A 734: 231
70. Boschetti E, Guerrier L, Girot P, Horvath J (1995) J. Chromatogr. B 664: 225

71. Roper DK, Lightfoot EN (1995) J. Chromatogr. A 702: 3
72. Thömmes J, Kula MR (1995) Biotechnol. Progr. 11: 357
73. Hall KR, Eagelton LC, Acrivos A, Vermeulen T (1966) I&EC Fundamentals 5: 212
74. Rowe PN (1975) Chem. Eng. Sci. 30: 7
75. Fan LS, Yang YC, Wen CY (1960) AIChEJ 6: 482
76. Veeraraghavan S, Fan LT (1989) Chem, Eng. Sci. 44: 2333
77. Bartels CR, Kleimann G, Korzun JN, Irish DB (1958) Chem. Eng. Progr. 54: 49
78. Belter PA, Cunningham FL, Chen JW (1973) Biotechnol. Bioeng. 15: 533
79. Gailliot FP, Gleason C, Wilson JJ, Zwarick J (1990) Biotechnol. Progr. 6: 370
80. Morton PH, Lyddiatt A (1992) Direct recovery of protein products from whole fermentation broths: A role for ion exchange adsorption in fluidized beds. In: Slater MJ (ed) Ion exchange advances. Elsevier, London
81. Morton P, Lyddiatt A (1994) Direct integration of protein recovery with productive fermentations. In: Pyle DL (ed) Separations for biotechnology 3, The Royal Society of Chemistry, London, p 329
82. Thömmes J, Born C, Biselli M, Wandrey C, Kula MR (1995) Purification of monoclonal antibodies by fluidized bed adsorption. In: Beuvery CE, Griffiths JB, Zeijlmaker WP (ed) Animal cell technology: developments towards the 21st century. Kluwer, p 515
83. Born C, Thömmes J, Biselli M, Wandrey C, Kula M-R (1996) Bioproc. Eng. 15: 21
84. Erickson JC, Finch JD, Greene DC (1994) Direct capture of recombinant proteins from animal cell culture media using a fluidized bed adsorber. In: Griffiths B, Spier RE, Berthold W (ed) Animal cell technology: products for today, prospects for tomorrow. Butterworth and Heinemann, Oxford, p 557
85. Vouté N (1992) Fluidized beds with beaded supports for protein separation. Proceedings of the second international conference on separation of biopharmeuticals, France
86. Somers W, Van't Reit K, Rozie H, Rombouts FM, Visser J (1989) Chem. Eng. J. 40: B7
87. Agosto M, Wang N-HL, Wankat PC (1993) Ind. Eng. Chem. Res. 32: 2058

Subject Index

Springer
and the
environment

At Springer we firmly believe that an international science publisher has a special obligation to the environment, and our corporate policies consistently reflect this conviction.
We also expect our business partners – paper mills, printers, packaging manufacturers, etc. – to commit themselves to using materials and production processes that do not harm the environment. The paper in this book is made from low- or no-chlorine pulp and is acid free, in conformance with international standards for paper permanency.

 Springer

Printing: Saladruck, Berlin
Binding: Buchbinderei Lüderitz & Bauer, Berlin